Robertson on Library Security and Disaster Planning

Robertson on Library Security and Disaster Planning

Guy Robertson

ELSEVIER

AMSTERDAM • BOSTON • CAMBRIDGE • HEIDELBERG
LONDON • NEW YORK • OXFORD • PARIS • SAN DIEGO
SAN FRANCISCO • SINGAPORE • SYDNEY • TOKYO
Chandos Publishing is an imprint of Elsevier

CP
CHANDOS
PUBLISHING

Chandos Publishing is an imprint of Elsevier
225 Wyman Street, Waltham, MA 02451, USA
Langford Lane, Kidlington, OX5 1GB, UK

The cover photo shows bomb damage that occurred during the Blitz to the Aston Webb façade of the Victoria & Albert Museum, Exhibition Road, London. Photo credit: Deborah Johnson.

British Library Cataloguing-in-Publication Data
A catalogue record for this book is available from the British Library

Library of Congress Cataloging-in-Publication Data
A catalog record for this book is available from the Library of Congress

ISBN 978-0-08-100077-9

For information on all Chandos Publishing publications
visit our website at http://store.elsevier.com/

Dedication

To my students.

Contents

Introduction

During their careers, librarians are inclined to whittle down their personal and professional goals into brief statements that summarize their philosophies and overarching purposes. Hence a public librarian in New York tells me that her "life's aim is to support democracy by providing people with the information they need." A children's librarian in Toronto says that her "heart's desire has been to bring joy to young people through books and reading." I much admire the medical reference librarian who wants "to improve human health and save lives by disseminating the most helpful information to physicians and researchers." A very original statement, and one of the more profound, comes from the director of a library in a large Asian museum. He swears that he serves as "the foundation to the foundation of culture."

That sounds like a heavy load, and I hope that he succeeds. If he does, he should not begrudge me a small amount of credit for his success, since my goal has been to keep libraries, their personnel, and their patrons safe. He and I have corresponded for decades, and our e-mail and letters have frequently included discussions of security and disaster planning. While he keeps me up to date regarding pandemics and tsunamis, I supply him with advice on preventing flooding and mold growth in his library. I have also warned him numerous times about the theft of items from his shelves, and about the risk of data loss from power outages, which occur often in his region.

In fact, he was one of the first people who encouraged me to write articles about security and disaster planning. He suggested that I write for a wide audience, be as clear and concise as possible, and try not to sound as if I were hungry for academic tenure. I should interview colleagues and record their opinions. While I should respect the theoretical aspects of my topics, I should always be practical. "And above all," he said, "try not to be dull." According to him, security problems and disasters should be "endlessly fascinating," and I should strive to make anything I wrote about them interesting as well as useful.

My editors gave me similar instructions, and added deadlines. Half of the articles were written during or shortly after events somewhere in North America: fires, severe weather, security breaches, or floods. I adopted an informal tone to make my articles more readable. Because damage to facilities and the loss of assets can lead to repercussions and finger-pointing, and cause much embarrassment, and since much of the blame is, in my experience, undeserved, I have respected the anonymity of most of my interviewees. I have changed their names, but quoted them as faithfully as possible.

It was tempting to "de-Canadianize" my writing in an attempt to make it more generally acceptable to readers—that is, to substitute Canadian place names with more familiar American or British place names. I worried that Kelowna, a charming city in southern British Columbia, would strike non-Canadians as "a little too far abroad." In the end, however, I retained the names of most Canadian cities, since I realized that they are no longer on the periphery of readers' mental maps. For a variety of reasons including tourism and immigration, Vancouver, Regina, Winnipeg, and Toronto are more widely known than ever before. (Kelowna is becoming more popular as well, in part because of the increasing quality of its wines. In future I should like to write a series of articles concerning risks to wineries—a fascinating topic.)

The articles in this book are divided into three sections. The first two sections contain articles that appeared originally in *Feliciter*, the magazine of the Canadian Library Association. The only exception is "People, paper, data," which appeared in *Disaster Recovery Journal*, a leading American trade magazine. The third section contains contributions to *Canadian Insurance*, which published a substantial amount of material concerning risk management and disaster planning, and *Canadian Bookseller*, which covered all aspects of the book trade including security issues and business resumption. I have included contributions to these magazines because I believe that librarians might benefit from a perspective slightly different from that which they find in library literature.

The articles in each section appear in the order in which they first appeared in print. Since I have been obliged to provide my readers with a theoretical context, there is a moderate amount of repetition from article to article, particularly when I rehearse the different phases of disaster planning: essentially what one should do before, during, and after a disaster. I hope, however, that there is enough variety in my approaches to specific topics to keep readers interested.

Perfection is impossible in preparing for any kind of emergency or disaster; it is equally impossible in journalism. I trust that readers will understand that these articles are more of starting point than a final answer to questions concerning the best ways to deal with negative events. This book is intended to serve as a sequel to my *Disaster Planning for Libraries*: *Process and Guidelines*, although it could also serve as a preparatory work or "prequel." Perhaps it is best to say that these books are sibling publications, and that readers might consider an examination of both. If not endlessly fascinating, they might at least be enjoyable and useful.

Guy Robertson
Vancouver (2015)

About the author

Guy Robertson lives in Vancouver. A graduate of the University of British Columbia's School of Librarianship, he teaches library history and records management at Langara College, and information management at the Justice Institute of BC. He has been an adjunct professor at the University of British Columbia's School of Library, Archival and Information Studies (SLAIS). He has lectured across North America on disaster planning, emergency management, corporate security, forgery, money laundering, and fraud prevention.

As a consultant, he has developed disaster plans and training programmes for libraries, archives, records centers, financial institutions, manufacturers, retail chains, hospitals, laboratories, seniors' residences, and architectural firms.

Other books by Guy Robertson

Disaster Planning for Libraries: Process and Guidelines

Acknowledgments

My gratitude extends to all of the many people who, over the past three decades, have encouraged and inspired me to write the articles collected in this book. Since most of these articles first appeared in *Feliciter*, the magazine of the Canadian Library Association (CLA), I should begin with heartfelt thanks to its editors: Liz Morton, Mary J. Moore, Peter Wilson, and the indefatigable Judy Green. On a number of occasions, copy-editor Rachel Hertz Cobb corrected my solecisms and saved me from embarrassment. CLA executive director Valoree McKay kindly arranged copyright permission.

CLA graphic designer Beverly Bard deserves a special note of appreciation for her efforts on my behalf. Beverly found ways for me to meet deadlines, most often by allowing me to stretch them by a day or two, or more.

I should like to thank *Canadian Insurance* editors Michael Steeler, Sally Praskey, Craig Harris, Barbara Aarsteinsen, and Stefanie MacDonald for their advice and assistance. I should also like to thank *Disaster Recovery Journal* editors David-Glen Smith and Jon Seals for their support.

At Langara College, I received advice and encouragement from Diane Thompson and Ryan Vernon. At the Justice Institute of British Columbia, Sarah Wareing provided encouragement when I needed it most. Her timing was impeccable.

The Elsevier Chandos editors George Knott, Harriet Clayton and Glyn Jones at the Kidlington (UK) office, Production Manager Preeta Kumaraguruparan from Chennai Office and Project Manager Kabilan Selvakumar from SPi Global deserves a special mention for their diplomacy and patience.

I am happy to acknowledge the help in different forms that I have received from friends, colleagues and former students: Ted Baker, Peter Broomhall, Virginia Carpio, John Livingstone Clark, Arthur Cohen, Jim Duggleby, Paul Evans, Heather Forbes, Bob Gignac, David Goldie, Drew Lane and Diane Guinn, Hilary Hannigan, Allen Higbee, Richard Hopkins, Rhonda Johnson, Steve Koerner, Kim Laudrum, Doug Little, Melany Lund, Lee and Teri Nicholas, Kelsey Ockert, Maureen Phillips, Stephen Porsche, the late Mahmoud Manzalaoui, David Mitchell, Teresa Murphy, Mike Rinneard, Judith Saltman, Dave Smith, Marguerite Stevenson, David Regher, the late Roy Stokes, Judy Thompson, Michael and Barbara Weston, Janet Whyte, Brigid Winter, and the late Anne Yandle.

As always, I am grateful to my wife Deborah Johnson and daughter Amanda Robertson for their patience, proofreading, and encouragement. I owe them another trip to Norfolk Square.

I am grateful to all who have contributed to my book. Any errors are my own.

Section A

Safety and security

Unofficial wisdom: A review of occupational health and safety

1

Across North America, occupational health and safety (OHS) committees strive to protect libraries from myriad risks. OHS committee chairs must struggle not only with fire hazards, poor air quality, and workplace violence but also with numerous misunderstandings that arise when people start thinking about life-safety measures and effective responses to dangerous circumstances. What follows is a memo from a committee chair on the verge of retirement to his successor. Under discussion are perennial issues that make the chair's job challenging. Originally confidential owing to the unguarded nature of its contents, this memo was leaked to the author by the usual reliable source.

To Whom It Must Concern: Our director has asked me to send you a memo regarding your new position as chair of the library's OHS committee. Of course you're surprised. You didn't know that this committee existed, and until now nobody has told you that you are to assume my old job. Congratulations. Take a deep breath, count to 10, and don't even think such vulgar things about our director, who is certain that you are the only person for the job. To be honest, nobody else wanted it.

By the time you receive this memo, I shall be lying on a Mexican beach, sipping a Corona and communing with my favorite Trollopes: Anthony and Joanna. I have retired, and none too soon. Some months ago it struck me that I had graduated from library school many years before the publication of AACR1. At times I felt as if I had graduated before the spread of books in codex form. It was time to go, and I did.

I have great confidence in the librarians who have taken over my duties. The reference and technical services departments will not suffer without me. (And, by God, I won't suffer without them.) But I was concerned about you. The OHS committee may not seem to be vital to library operations, but it is. It took me months to realize that an OHS program is necessary for the ongoing health and morale of library staff and for the general satisfaction of patrons. Many people are unaware of what an OHS committee does, because its success lies in what doesn't happen: the injuries that don't occur, the building components that don't break down, the bad news that doesn't circulate. What follows are 20 OHS items and issues to consider before you chair your first committee meeting.

1.1 The myth of the omnipotent committee

Do not assume that you and your committee will be able to implement without delay all of your excellent ideas about staff safety and workplace improvement. You may have a plethora of guidelines from the workers' compensation board, the local fire

Robertson on Library Security and Disaster Planning

department, the Ministry of Health, and city hall, but it usually takes longer than you expect to improve the lighting in the underground parking lot and the ventilation in the staff room. In fact, it takes too much time to get anything done, and unless something is about to explode and the library is on the verge of vaporization, OHS tasks are usually drawn out, postponed, put on hold, tabled, or taken under advisement. (In Mexico, they say *mañana*, which means "tomorrow" or "never." Library directors and boards are familiar with both usages and prefer the latter when considering the matter of antislip doormats and ergonomic workstations.)

1.2 The myth of the omnipotent director

She can't do everything. She's your director, and everybody wants her time. Department heads, branch heads, systems managers, and board members demand her attention now, ASAP, and at her earliest possible inconvenience. Inevitably she's perceived as the ultimate doer, the person who controls the library. When she does what people ask, she's a top-notch manager and leader. When she doesn't, she's irresponsible, out of her depth, and unprofessional. Poor soul, she's only human.

On any given work day, she must contend with a shrinking budget, an angry union, three outraged members of the public, and a journalist who wants to know why she allows little boys to view Internet porn in the children's department. Just as she's thinking of a new way to tell the mayor that she's nervous about renting the public presentation room to the local neo-Nazi chapter, you arrive to express your concerns about loose handrails in the stairwells.

Your director will help you as best she can, but don't expect miracles. OHS is but one of her ongoing worries.

1.3 The Paper God

Paper is the bureaucrat's strength, the communicator's hurdle. I refer to all the policy and procedures manuals, bulletins, newsletters, and memos that we produce to get our point across. Frankly, most OHS material is ignored or forgotten as soon as it is out of sight. Those instructions concerning emergency evacuation may make good sense, but most people who bother to read them assume that they will never have to leave the library in a hurry unless *Coronation Street* moves to an earlier time slot. Safety is something that's taken for granted. After all, genuine emergencies are rare. So there's really no need to study that three-ring binder full of OHS advice…

But as chair of the OHS committee, you recognize that there are serious reasons for every library employee to know the contents of that binder. To spread the word, you must defy the Paper God and skinny-down the message. Instead of circulating the binder, send out a two-panel brochure with the pay stubs. You'll find that most essential life-safety advice can be distilled into a relatively small format. And you can use brief, point-form e-mail messages to remind staff about OHS issues. But try not to sound like a mortician with a hangover when you compose OHS material. Feel free to use cartoons and other forms of humor. Above all, do not sound like a tort lawyer, a sociologist, or (worst of all) a library educator when speaking or writing about OHS.

1.4 Mythical beast number one: The disaster plan

The Paper God exults when he hears that you have been given the task of compiling the library's disaster plan. Now you can spend the next decade developing another chunk of bumf regarding the correct attitude to all manner of dreadful events, including natural risks, such as floods, high winds, severe weather, earthquakes, and fires, and human-caused risks such as arson and sabotage. In fact, the library already has a disaster plan, but nobody knows where it is. I suspect that it's in the same place as all the other disaster plans that the OHS committee has produced over the years.

As soon as any disaster plan is produced, either by the committee or by a team of consultants, it is studied by the director, passed on to the board, approved in glowing terms, and passed back to the director, who gives it to someone else, who promptly loses it. The chunkier the plan, the faster it will reach that final destination in the janitor's closet of an outlying branch, or in a box containing politically incorrect Christmas decorations.

If you succeed in forcing library staff to take the disaster-planning process seriously, you will have accomplished a miracle. Just remember that the plan comprises three separate but related programs that cover your circumstances *before* all Hell breaks loose, *during* the actual outbreak of Hell, and *after* the devils have done their worst.

Preparedness is the theme of the first program, which involves the discovery and mitigation of risks that prevail at the library. Response is the theme of the second: that is, what you do to save your neck when the library catches fire (Get out!) or an earthquake strikes (Stay in!), or an ice storm causes a power outage and makes commuting to the workplace a long and dangerous struggle (Stay home!). Recovery and service-resumption guidelines constitute the third program, which can be the most time consuming.

You needn't produce a chunky disaster plan. For a public library with 10 sites (one garish monstrosity in the city center and nine quaint branches), your plan should be no more than 60 pages. Of course you'll be obliged to include all sorts of appendices, some of which are actually useful. But the core of the disaster plan should be as concise as possible.

Remember that you should audit the plan annually so that it will reflect your current circumstances: the new branch, new staff, or new system. As for the natural risks, they remain constant. Once a tornado zone, always a tornado zone.

1.5 Mythical beast number two: The security plan

More paper. But you can reduce your security plan to under 20 pages in many cases. The easiest way to handle this beast is to survey the staff regarding what they perceive to be security risks. They'll tell you about the substandard lighting in the parking lot, problem patrons at a branch, drug deals in the men's room, and suspicious characters loitering in the children's department. You should take all of these observations seriously.

The security plan covers all security risks as concisely as possible. A good general rule for all staff: stay out of harm's way. This sounds commonsensical, but you'd be amazed how often librarians insist upon dealing with people and incidents that are best left to the police. You shouldn't try to reason with a knife-wielding addict or threaten to remove his borrowing privileges. The emergency number is 911: encourage library staff to use it.

Security risks change over time as a suburb becomes part of the inner city and the old branch attracts new and occasionally rough users. You're wise to audit your security plan annually, more often if there are increasing reports of security breaches such as vandalism, attempted thefts, and break-ins.

1.6 Mythical beast number three: The workplace violence program

Numerous governments are enacting legislations to deal with workplace violence, which is usually defined as physical violence in a workplace or the threat of physical violence. While there are few statistics collected on workplace violence in North American libraries, it is reasonable to assume that the incidence of physical violence directed at library employees is lower than that observed in some workplaces. Still, we are well advised to be on our guard, particularly in large cities.

Most workplace violence programs involve a staff survey and the formulation of a policy—yes, more paper. But you can combine your workplace violence program with the security plan and satisfy the legal requirements in most provinces.

1.7 Mythical beast number four: Accusations of alarmism

Many people will assume that all is peachy at their branches: then you and your wretched OHS committee arrive with survey forms and safety tips and… "Aren't there more important things to do around here? Come on, this isn't New York City or LA. Why do we need another committee to handle incidents that will probably never happen? Hey, you wanna solve a problem for me? The system was down for 3 days last week and I'm short-staffed and the amount of manual input is growing and you come around and start asking about workplace violence. You go back to Central Branch right now and tell that systems manager that there's going to be bloodshed unless he does something about my blah blah blah! Maybe you should reexamine this library's priorities before you start wasting my time and every other branch head's with another three-ring binder full of material that I won't look at, I promise you."

It was ever thus. Being accused of alarmism is part of your job, and you must get used to it—until something awful happens, at which time you become either a hero whose foresight saved the library, or an incompetent who could not persuade the Branch Head quoted above to take the necessary precautions.

1.8 Mythical beast number five: Apathy

Who cares?

No one will ever say this at your face, but you can expect many library employees to think this way. Apathy underlies almost all of the different tactics used to delay the OHS committee's work. Accusations of alarmism, willful ignorance, and declarations of higher priorities can thwart you, but you must persevere. Take heart: at least some staff members will appreciate your efforts.

1.9 The myth of perfectable heating, ventilation, and air conditioning

Is there a library that can boast a perfectly satisfactory heating, ventilation, and air conditioning (HVAC) system? Maybe in Heaven, never on Earth. You hear the same comments in libraries all over the world: "The air on this floor is really bad" and "It's freezing in this branch every summer, and boiling in the winter" and "We can barely breathe owing to the humidity in this staff room." If only the board would agree to spend next year's entire budget on upgrading or replacing the current HVAC system. Then the air would be pure and unvarying in temperature, and all would be well. Staff members wouldn't catch cold or flu, and allergies would disappear.... Not likely.

In fact, pure air exists rarely outside of certain laboratories; what we breathe usually includes contaminants. Fortunately, our lungs can filter out many harmful agents. But still, some workplaces are uncomfortable and possibly unsafe because of substandard ventilation or unreliable heating. Here you must be vigilant and keep records of the problem, including the times during which it is most obvious, specific locations, and any outstanding features such as a noise coming from HVAC machinery or a distinct odor from the vents. The property manager should receive copies of these records as well as reminders that the problem should be solved or at least controlled. In most cases, a HVAC technician can adjust the system so that air quality is bearable.

Remember that it's not the HVAC system that the OHS committee is obliged to deal with. Rather, you must concentrate on the property manager. If weekly requests for air quality improvement do not motivate him, try daily requests. Then hourly. If the property manager won't respond to the latter, threaten to organize a picket line outside his office. Once upon a time in the U.S. Midwest, librarians at a college library set up a protest picket outside the office of their director of facilities, who had refused to schedule repairs to the library's ventilation system. He capitulated when the protesters started to chant: "Breathe in, breathe out, one, two, three; clean air means a lot to me!"

Don't laugh. It worked.

1.10 Death, taxes, and leaky plumbing

Like everything else, plumbing ages. It cannot last forever. Occasionally it ruptures, causing a leak or a flood. To avoid plumbing problems, ask the property manager to inspect pipes, sinks and other washroom fixtures regularly. If a rupture seems imminent in any area of the library, repairs should be made at once. Suspicious stains on walls and ceilings should be investigated: these can result from a break in the building envelope as well as from a broken plumbing.

If there is any way that you can persuade branch heads and supervisors to learn how to turn off the water in their buildings, do so. Often the property manager and repair personnel cannot arrive at the building as soon as they're needed, and further water damage will occur.

Fortunately, leaks are seldom life-threatening, but water in contact with electrical wiring increases the risk of fire, and water on floors can result in slip-and-fall injuries. Preventive maintenance (an OHS panacea) is advisable for all plumbing systems, particularly those in older buildings.

1.11 The mystery of signage

We believe in signage. Religiously. It strikes us as a form of continual communication. A good sign is unambiguous: everyone will understand its message instantly. There is only one drawback. Just because a person can read and understand a sign does not mean that he or she will take it seriously. Often users will seek confirmation of a sign's message. For example, the reference librarian sits under an enormous reference sign, she wears a librarian identity badge, and facing outward on her desk is a sign that indicates that she is indeed a librarian. How often do people approach her and ask, "Are you a librarian? Can I ask you a question?" This is a common human response to an unambiguous message.

Librarians respond similarly to safety signage. A fire breaks out in the staff room; the librarian runs to the fire alarm. He reads the little sign above it: "In case of fire, break glass." He runs to his branch head to ask if it's permissible to break the glass in order to activate the alarm. The branch head considers the question carefully. She is not sure. Nor is she sure about dialing 911. After all, this involves a fire, not a heart attack or an armed robber. The sign above the phone says *Emergency*: *Dial 911*. Breaking the glass should be permissible. Or dialing 911. But both?

It's getting hot in there. The Exit sign has melted. We believe in signage, but often we don't take it seriously. Why not? It's a mystery.

1.12 Vertebral follies

Which is heavier, a ton of iron or a ton of science fiction? Library staff members will tell you that they could never lift a ton of iron, but they might be tempted to do something foolish with science fiction. Or biographies. Or CDs. In fact, library employees are notorious for lifting heavy materials in ways that are almost guaranteed to cause back problems—if not immediately, then at some point in the future. In any library you'll see staffers bending at the waist to pick up crates of books. Signage posted nearby tells them to bend at the knee. Will they heed this advice? See "The mystery of signage" above.

You can reduce the risk of back injuries by encouraging supervisors to warn staff away from dangerous lifting techniques. Staffers may ignore signs, but they will pay attention to "Yo! Bend your knees, or you'll wreck your back!"

1.13 The myth of the magical first aid kit

The first aid kit is useless unless there is somebody present who has been trained in first aid. Nevertheless, many librarians believe that the presence of the first aid kit wards off danger, just as the Hollywood request for boiling water and towels signals the imminent birth of a healthy baby. (Have you ever seen a movie in which towels and hot water did not lead to the arrival of a healthy screaming baby?)

First aid kits are not magical. Encourage library staffers to take a first aid course. Take one yourself, and lead by example.

1.14 Obsessive focus

Too often, OHS committees become focused on one risk in particular: poor lighting in the staff room, or creaky, unsound chairs in a public area, or the number of asthma cases in the fine arts department. The Paper God gets busy; the memos flow like a river in spate. Countless hours are spent in discussing possible solutions to the problem. But the real problem, which can lead to serious results, is the obsessive focus on the lighting or those chairs or that asthma. Most libraries have a number of OHS problems that require your consideration. Don't get hung up on one. To narrow your risk exposure, keep your perspective broad.

1.15 Men, their strength; women, their fragility; myths, their tenacity

This is a sensitive issue that requires the utmost diplomacy, except in a memo such as this, which contains nothing but the unvarnished truth.

Tell me, do you know a library staffer who can bench press 90 kg? Who can run 15 km without stopping for a rest? Who is in top physical condition? I am acquainted with one library technician and three librarians who fit this description. They're all women. There are probably a number of male library staffers who are in good shape as well, but I haven't met one. Despite my enormous vanity, I can't include myself; despite my unfailing generosity, sir, I can't include you. Face it: like most of our colleagues, we are not physically fit.

Librarianship is a sedentary profession, and so our inability to pump large amounts of iron or run a few kilometers without a rest should come as no surprise. As chair of the OHS Committee, you should remember that most staffers are out of shape, although there are exceptions of both sexes. For the purposes of leading an evacuation, you are prudent to rely on somebody who is fit and strong. It's unlikely that anyone will accuse you of discrimination during an emergency, since the need for physical strength under such circumstances is obvious.

Have I offended you? You can sue me as soon as we're outside this burning building.

1.16 The myth of the magical fire extinguisher

Nero fiddled while Rome burned. In fact, he was fiddling with the instructions for his fire extinguisher. He hadn't bothered to read them before the fire broke out, and look what happened.

There is no point in mounting a fire extinguisher in an obvious location unless there is somebody on site who knows how to use it. It will not function magically on its own. Consult the local fire department for information on extinguisher training, which is not only useful but also a lot of fun.

(Incidentally, have you signed up yet for that first aid course? Nag, nag, nag.)

1.17 The bomb that didn't go boom

Your OHS committee should regularly enunciate its policy regarding the appropriate response to bomb threats, all of which must be taken seriously. Often the search for a bomb planted in library turns up nothing or a phony bomb that contains no explosive. It is unwise to assume that all bombs will be equally harmless. The OHS committee should consult the local police for advice on dealing with bomb threats. The appropriate procedures should be included in the security plan.

1.18 Legends of causal folly

John smokes three packages of cigarettes a day. He complains of the poor air quality in the children's department where he works. He suggests that his breathing problems and allergies are due to that bad air. Do not giggle when he tells you this, for ironically he might be correct. There is a chance that poor air quality is causing his problems or at least exacerbating them. It is far more likely that his smoking is the cause, but you are not a physician and you do not have the training or diagnostic tools to determine the truth in this matter. All you can do is take John's complaint seriously and investigate it.

Jane is 40 kg overweight. Her posture is frightful. You observe her bending at the waist to lift large boxes and equipment. She suggests that her sore back is due to the shape of her chair. She wants a new and very expensive ergonomic chair to slouch on. Again, no giggles. Her sore back might be caused by her old chair. It is not your business to examine her X-rays or recommend a brace. You might have to provide her with a different chair, even if it isn't ergonomically designed. Aside from that, all you can do is to advise her to be careful when she lifts heavy things.

1.19 The perils of public washrooms

Some day, a librarian will write a book about washroom security, and it will become a best-selling work of library literature. Why? Because so many public and academic libraries have washroom security problems. OHS and security specialists talk about the three Fs of washroom crime: exchange of fluids (e.g., dope), funds (cash for dope), and fun (hello, sailor!). While the men's room is the most common scene of library crime, the women's room is increasingly popular.

No public washroom will be completely secure, but there are ways to make one less risky. Some libraries ask security staff to patrol washrooms regularly; branches often insist on holding washroom keys at the circulation desk until a user asks for one. In order to respect users' privacy, no security cameras can be mounted in a washroom.

Incidentally, no library washroom should be used for the storage of paint, cleaning fluids, and other chemicals. These highly toxic and flammable items should be locked away in properly ventilated closets.

1.20 The myth of the OHS committee that finished its job

Forget it. Since risks never disappear altogether, neither will the job of your committee. There will always be another task to complete, another OHS survey to review. Occasionally you will feel as if you're wasting time. You won't see the point of your efforts, since security breaches and accidents are so infrequent. Remember that it could be the work of your committee that is making them infrequent. As for natural disasters, you may not be able to prevent them, but at least you can mitigate their worst consequences.

Your first committee meeting begins presently. The agenda includes slip-and-fall injuries in the parking lot, the purchase of ergonomic chairs for the technical services department, carpal tunnel syndrome in the circulation area, and the disaster plan, which is overdue for an audit. Best of luck and *adios*!

Our vision of things: Basic eye care for librarians

2

As children, we played the blind game. Walking along with Mum, clutching her hand for security, we closed our eyes and pretended that we were completely sightless. The pavement felt harder under our feet, and street noises sounded louder and slightly threatening. At last we could endure our self-imposed darkness no more. Opening our eyes, we reclaimed the entire world; and of course it was a relief to sidestep the muck or puddle that lay only a stumble ahead.

Early on, we learned the value of sight, upon which we have depended on so much for our education and professional development. Most librarians realize that of all senses, they use their sight more than any other in their jobs. But the library is no haven for the eyes: in fact, we spend much of our careers in an environment that seems almost designed to cause eye problems.

We begin with library school. Look at those old class photos, which usually trigger so many memories and expressions of wonder. (How did *she* become a director? She barely made it through first-term cataloging, and then only with *my* notes. And look at *him*! In those days he had hair. And there's my long-lost colleague…) But consider how many classmates wore glasses. Remember those long nights squinting over the tiny print in directories and bibliographies in the dim light of the university reference section? Or the research project conducted in a dusty storage unit where the air tasted of rotten leather or mold? Or the way the world turned into a blur after 4 hours in front of a reading room terminal? It's no surprise that so many library students need a prescription for stronger lenses before they graduate.

The ocular challenge becomes more intense when we arrive at our first professional workplace. In many Canadian libraries, ergonomic policies cover little more than chairs and lifting techniques, with almost no attention to working conditions that jeopardize eye health. We treat the symptoms of eye disease as they arise. Headaches related to eye strain make us reach for that aspirin bottle (in the circ. desk drawer beside the Tums); red, itching eyes lead to mirror inspections in the staff washroom followed by a dash of Visine and a promise to "do something about those contacts." Just because the cause of an eye problem is obvious does not mean that librarians will attempt to mitigate it. Consider the terminal from Hell on the reference desk: staff members have complained about it for years. Its screen delivers a blast of glare that makes us sentimental about the card catalog. But usually we allow that electronic brute to have its way; it won't be removed until it burns out.

The *Concise Oxford Dictionary* defines glare as "strong fierce light." In libraries, glare emanates from several sources apart from the terminal screen. Glossy magazine pages are often difficult to read owing to the glare that they cast. Lighting systems may illuminate passageways effectively, but they make reading an uncomfortable experience. Patrons can complain about the lighting, check out their books, and go home to glareless surroundings; librarians must work under those shining tubes (that also hum and crackle: see Aural nuisance) for years.

Another cause of eye problems is the physical atmosphere inside the library. It is rare to find a library building that enjoys a trouble-free and efficient heating, ventilation, and air conditioning system. Many library spaces are constantly hot and dry. Since book collections become dusty and dust particles are stirred up throughout the shelf area, the library's air filters can be overwhelmed. The mixture of dust and dryness is hard on the eyes and can lead to eye problems such as allergic reactions.

How can we maintain good eye health and protect ourselves from glare, dust, and fine print? The following tips are so commonsensical that most of us have ignored them since we entered library school. It's time to reconsider them seriously, since library budgets are shrinking and our workplace conditions are not likely to improve in the near future.

2.1 The maintenance of good general health

Doctors say that they can tell a great deal about a person's general health simply by examining his eyes. Diabetes, hepatitis, and vascular disease can affect the eye, but an unhealthy lifestyle also contributes to many eye problems. You've heard it all before. Eat less, exercise more. Stop smoking, reduce alcohol consumption. Get more sleep. Learn to control stress. The simpler the advice, the harder it is to follow. But you don't have to become a saint overnight, and you can adapt gradually to a better diet or a regular exercise program. Good general health is the foundation of good eye health.

2.2 Regular eye examinations

We should have our eyes examined by an ophthalmologist or an optometrist at least every 2 years. Those with a history of eye problems should be wise to have an annual check-up.

Eye examinations begin with an inspection of the face, eyelids, and eyes, followed by a visual acuity test based on the Snellen Chart, which features that big E at the top. Both eyes are tested independently to determine the sharpness of a person's vision, and the results are expressed in ratios such as 20/40, which means that at 20 ft., that person sees what someone with normal vision would see at 40 ft. The higher the second number, the worse the person's vision.

The eye examination will also include inspections of the pupil and retina and a test for glaucoma, a disease involving an increase of pressure inside the eye. If a person is 65 or older, she will be tested for other conditions including cataracts. It is wise to tell the eye examiner about any recent changes in visual acuity and any problems related to eyeglasses and contact lenses.

2.3 The right corrective lenses

An eye examination will determine whether a person's glasses are appropriate for his different activities. It is not unusual for a librarian to have a dedicated set of eyeglasses for use when he is working at a computer terminal for extended periods. That librarian might also have another pair for reading and another for distance vision. Remember that our vision changes over the years, and new lenses might be necessary more often than we expect.

The eye examiner might want to look at the frames of our eyeglasses to make sure that they do not irritate the areas they come into contact with on the face. Older frames are sometimes too heavy and should be replaced with newer, lighter models. Are you still wearing the frames you wore when you studied AACR1? It's probably time for a change.

2.4 Proper hygiene and first aid

The rule of thumb: keep the thumb away from the eyes. Keep the fingers away from them, too. While a library is cleaner than many workplaces, its dust can irritate the eyes, especially when a librarian insists on scratching that itchy eyelid with fingers that have just held a frequently handled reference work or periodical. Fingers transfer dust and microbes directly onto the surface of the eye, and irritations and infections can ensue. Regular hand-washing with soap is a good preventive practice against a number of health hazards, particularly for those who suspect that they scratch the area of their eyes unconsciously.

Some people become so entranced with what they are reading that they neglect to blink. The result is dry eyes, which quickly become irritated. Nonblinking text inspection seems to be more common among computer users than readers of hard copy. The best remedy is our own tears, which are not as idle as Tennyson assumed.

Eye emergencies in libraries include severe allergic reactions, blows from falling objects, foreign objects in the eye, and paper cuts. Library first aid attendants should advise a person with an eye injury to seek medical treatment at once—if not in a doctor's office, then at the nearest hospital emergency ward.

2.5 Enlarging fine print: Elementary, my dear...

Once upon a time most reference desks contained a large magnifying glass for use on those hardcopy abstracts that could be accommodated on the head of a pin. There is still an enormous amount of fine print in our collections; there is ample justification for that magnifying glass. Get one, and use it. Do not assume that squinting improves visual acuity.

2.6 Let there be the right level of light

Our computer workstations are often surrounded by too much light, which washes out the image on the screen and forces our eyes to do more work than they should. In technical services areas, it might be possible to lower the lighting, but in reference sections and other public areas lighting must remain bright for the safety and convenience of patrons. In these circumstances a hood can be added to the terminal to decrease washout and enhance screen images.

Antiglare filters can be placed over screens to decrease eyestrain. These filters are inexpensive and effective. Screen flickering can be reduced by adjusting the brightness control.

2.7 The issue of color

For most computer users, green or amber screens are more comfortable to use than black and white. This does not mean, however, that you can use a green screen for unusually long periods. At some point, any screen will tire the user.

2.8 Simple workstation ergonomics

The chair is elegant hardwood from the early 1900s. It does not provide proper back support. It cannot be adjusted and offers no cushion. The computer is based on a high table that shoves the screen onto the user's face and offers no forearm and wrist support. This workstation is a recipe not only for eye fatigue but also for back problems and repetitive strain injuries. It will decrease productivity and contribute to absenteeism. It is a standard equipment in many libraries, and it should be replaced as soon as the budget allows.

2.9 Ocular R & R

The myth is that because librarians do not appear to be moving energetically at all times, they are not working hard. Librarianship is a soft option, unlike the building trades and jobs in heavy industry. The unfortunate aspect of such nonsense is that many people—including some who sit on library boards—believe it. In fact, the eye is controlled by six muscles that will tire with extended use. When our eyes get tired, usually we are in need of a break. Ocular R & R should include freedom from print, computer screens, and bright lights. It can involve a snack in a comfortably lit staff room or a spell of meditation while gazing out a window into the distance. It should not involve newspapers or computer games.

Incidentally, the muscle bound construction worker will fade as fast as anyone else in a technical services area. The muscles in his eyes are not necessarily stronger than those of the unathletic cataloger. After all, you do not pump iron with your eye muscles.

2.10 Eye exercises

There are effective ways to exercise our eyes and strengthen the muscles that control them. These exercises often involve focusing on objects at different distances from the eye. It is best to organize a personal eye exercise program with the advice of an ophthalmologist or optometrist, especially if we have corrected vision or a history of eye disease. It is also essential not to overdo eye exercise. We cannot strengthen our eye muscles in such a way that we can improve our vision to super-human levels. Moderation is the best policy.

Librarianship involves much eye-straining work. We are obliged to protect our eyes as best we can under circumstances that are difficult to control. With a modicum of common sense and effort, however, a high standard of eye care is feasible for every librarian. Perhaps the best motivator is the blind game: try it on your way home this evening. Just remember that the next curb is closer than you think.

Stress? Settle for less

3

Take this simple test. Ready? Read on.

"So what do you do in your library, anyway? Must be pretty slack. I mean, shelving books is not exactly rocket science. You tell stories to toddlers? Wow. That's tough. No, just kidding. But let's face it, you're not killing yourself as a librarian."

Finished? Good. Check your pulse rate. Is it increasing rapidly? If so, you're a librarian.

At this point you're to be reminded that tearing up professional literature is against the regulations and punishable by another festive dinner with the same in-law or bachelor uncle who insists on telling you how lazy librarians are. Where does he get that idea? In fact, the stereotype of the hair-in-a-bun-and-glasses librarian hides an even deeper misconception of her work, which is that it's not particularly hard and that it doesn't involve stress.

This misconception contributes to the life insurance industry's belief that librarians are secure, unstressed, and low-risk, and thus eminently insurable. Meanwhile, the authors of self-help books promote librarianship as a gentle alternative to the more demanding professions. In *Beating Stress at Work*, Anne Woodham offered "librarian" as an example of a low-stress job, along with museum curator, gardener, farm laborer, craftsman, and computer programer.

In candid moments, library school directors recollect applications from unfortunate persons who have been steered in the direction of library work by psychiatrists and other mental health consultants. Letters of reference arrive, stating that while poor Ms. Doe has had several breakdowns and a long history of depression, she needs a job and is cut out for a relaxing position in a library. (Ms. Doe is allergic to dust: a senior manager's job away from those shelves would probably be best.) Directors are not inclined to talk about how they treat such applications, although it's safe to assume that certain psychiatrists have received some brusque replies to their letters of reference.

To be fair, there are low-stress positions in libraries. We are obliged to complete numerous routine tasks that require little effort or intelligence. In fact, all professions involve such tasks. Doctors are hardly challenged by most of the ailments that we bring to them; the expression "it's not brain surgery" can be traced to the attitude that doctors have toward wart removal and the diagnosis of the common cold. Lawyers feel the same about conveyancing and simple wills; architects and engineers don't fret about straightforward drafting.

Even in the absence of scientific surveys, however, it is obvious that librarians are suffering heightened levels of stress. Your workload may be light, but you can't relax: after all, tomorrow you might be considered redundant and laid off. Or your workload may be heavy and your responsibilities crucial to the ongoing quality of your library's services, but the mandarins in city hall don't care, because their consultants have charged them a high fee for a report that suggests that your library can do without you.

Robertson on Library Security and Disaster Planning

Stress arising from fears of job loss is endemic not only in public and academic libraries but also in many special libraries where the librarian can be replaced by the technician, the technician by the clerk, and eventually the clerk by the temp from the mail room. In a time of downsizing (certainly the ugliest term from the 1980s) and increasing unemployment, none of us can rest easy, and all of us experience more stress than we need.

Other recent causes of stress include the implementation of new library technologies. Our response to these technologies leads to what Craig Brod describes in *Technostress: The Human Cost of Computer Revolution* as the feeling that we are quickly falling behind in our understanding of the uses and operation of new systems, and that our main value as professionals springs from our systems expertise. The Luddites among us have become objects of pity and scorn: that librarian knows nothing about the magnificent new database and should be let go as soon as the human resources manager can print out a pink slip. The future is electronic, and soon our patrons will refuse to accept a measly hardcover book. They will demand Jane Austen in Hypertext, Dickens on CD—that is, if they bother with fiction at all. Where these attitudes prevail unchallenged, they create a stressful atmosphere, particularly for librarians who question the meaning and effects of the Digital Revolution.

By the way, are you completely up to date? Do you have an intimate knowledge of metadata and RDA? If not, why not? Remember that there are hundreds of fresh library school grads who are eager to replace you if you can't adapt to your library's new system. Those who control your library's budget are aware that it's an employer's market and that you can be replaced by someone more computer literate at a lower cost. *O tempora, O mores*: what can you expect?

How should librarians deal with stress? First, it is essential to realize that we cannot eradicate stress from our lives. In the end, we wouldn't enjoy stress-free living, since it quickly becomes dull. Moreover, there is such a thing as positive stress, which is referred to as "eustress." An example is the pleasurable tension we feel in our efforts to complete a project that fascinates us. Many of what we will eventually consider personal achievements are in part the products of eustress, which acts as a motivator.

As for negative stress, in most cases we can only reduce it to tolerable levels. One mistake many of us make as we begin a personal stress-reduction program is to focus on only one source of stress (or stressor) and neglect others. For example, Jane runs a technical services department in a mid-sized public library. During the implementation of a new automated system, she begins to experience panic attacks and heart palpitations. She craves food and begins to smoke heavily. Her colds last longer, and she has difficulty falling asleep.

It is easy to identify these as classic symptoms of stress, but the new system is not necessarily their main cause. Jane is also in debt. She has recently married and is anxious about child-bearing, making ends meet, and her new in-laws. In fact, Jane is affected by multiple stressors. Her concentration on only one might be a way to avoid dealing with others that are more difficult. After all, that new system could be a minor concern when compared with a horde of new in-laws.

The myriad books on stress related topics contain similar advice regarding the most effective ways to reduce the impact of stress. Adapted to meet the needs of librarians, they can be summarized as follows.

3.1 Improve your working environment

- Make sure that your library's heating, ventilation, and air conditioning (HVAC) system filters the air you breathe and maintains reasonable temperatures. In many regions, HVAC maintenance is a constant challenge owing to extremes of climate and the age of many library buildings. But acceptable levels of heating and filtration can be achieved through regular maintenance. Don't allow your library's HVAC to fall into disrepair. If you're uncomfortable with the temperature or the air you're forced to breathe, you have every right to complain.
- Make sure that lighting levels are appropriate to library activities. Most libraries don't need high-intensity lighting. Find ways to take advantage of natural light. Must those curtains and blinds be drawn at all times?
- Decrease the noise level. Loud noises are infrequent in libraries, but there can be background noise from photocopiers, computers, and telephones. If possible, adjust the telephone bell to a lower level and move photocopiers away from your immediate work area. If the hum of your computer is distracting, it might require servicing. In some libraries, the computer need not be turned on at all times.
- Make your workstation more ergonomic. Be sure that your chair is adjusted to the most comfortable height and that you have adequate support for your arms as you work on the keyboard. If your chair does not provide adequate support for your lower back, either replace the chair or use firm cushions. Invest in an antiglare filter if your terminal is hard on your eyes, and adjust the brightness of your screen to avoid flicker.
- Personalize your work area. Put a picture of your spouse, partner, child, dog, or cat on your desk. Cards, flowers, and posters are also appropriate. (Incidentally, why does anyone feel the need to hang a degree on the wall? One's child is unique; one's MLIS is not. The memories of library school generated by the sight of that degree are hardly stress-reducing.)

3.2 Use your working hours to your advantage

- Try to avoid long stretches of overtime. Systems librarians are notorious for putting in too many hours, especially during conversion projects. Eventually this schedule can prove counter-productive, since it often leads to more sick time and absenteeism.
- Take a break. Your lunch hour is for a sandwich, a cup of tea, and 20 pages of that bad novel you don't want to put down. Don't use this time to tally your reference stats, survey the branches circulation figures, or attend another committee meeting.
- Organize your work so that you don't have to take it home. Why clutter your personal life with another wretched report? Is it really necessary to neglect your family for the sake of a user survey? Most evenings and weekends should be devoted to family and personal interests.

3.3 Make commuting as pleasurable as possible

- If you travel to work by car, listen to music that you enjoy. Listen to comedy tapes and audio novels. Avoid using mobile phones while driving. Don't try to compose a memo on your laptop while waiting for a light to turn green.

- If you are involved in a car pool, interact with the people with whom you travel. Don't put on the headphones and ignore your companions. Treat the pool as a potentially pleasant social gathering, and talk about gardening, pet care, or football—but not the revisions to the Library of Congress Subject Headings or the latest news regarding RDA.
- If you travel by bus or subway, you can use those headphones to enjoy a morning or afternoon dose of Schubert, Brahms, or Lady Gaga. Or you can follow the British example and work on a newspaper crossword. Or you can meditate.

3.4 Enhance your professional development

- Just because you have completed your Master's degree doesn't mean that you've finished your education. If you are worried about losing your job, investigate new occupations and training programs and get involved with something that not only makes you more attractive (and thus less expendable) to your current employer but also attractive to other employers. Many librarians have taken courses in ESL training, computer hardware maintenance, early childhood education, geriatric studies, recreation management, bookkeeping, interior design and decoration, records management, travel writing, and fashion merchandising. Librarians who receive additional training in different fields say that they are less worried about job loss and more optimistic about their futures.
- If you're not computer literate and want to be one, you can usually find a course or training program that will give you the background that you need. Remember that all of us were utterly ignorant of systems routines to begin with and that everyone learns at a different pace. You shouldn't feel guilty about your ignorance of a particular software package. You can always blame the vendor for failing to give you—the esteemed user—adequate orientation and training. Picking on Bill Gates is a guaranteed stress-buster.

3.5 Improve your health and physical conditioning

- Experts on stress reduction agree that improved diet, exercise, and sleep are key elements of stress reduction. This does not mean that you must immediately lose 20 pounds, take up aerobics, and chain yourself to your bed for at least 8 hours a night. Despite all the advice in self-help books and popular magazines, you should avoid fad diets. You should try to find a form of exercise that suits your needs, and you should not feel obliged to take up jogging, cross-country skiing, or weight lifting. Sleep patterns vary from person to person, and some of us need far less than others. It is important to be aware of how much sleep you need and get that sleep every night if possible.
- Have regular medical check-ups. If you believe that stress is having a negative effect on your life, discuss it with your doctor. It is prudent to tell your doctor about any plans you have to change your diet or take up a new regimen.
- Respect the obvious. Stop smoking. Don't drink too much liquor or caffeine. You've heard this advice a thousand times. Now take it seriously.

3.6 Increase enjoyable family time

- Date your spouse or partner. Go ahead: surprise him with a phone call and ask him out. Chances are that he will accept. When he does, agree that neither of you will discuss any aspect of your work over that candle-lit dinner.

- Spend time with your child, but don't turn every parent and offspring meeting into a Gestalt session. Engage in activities that you can both enjoy. Remember that much of the stress of childrearing arises from what you don't know about how your child feels and thinks. When your child talks, listen. And don't turn that bedtime story into a professional occasion: you're reading to your flesh and blood, not a group of storytime attendees.
- Ask your parents to tell you about themselves. There is a good deal that you don't know about your mother and father. Give them an opportunity to inform you. You might be amazed; you will certainly be entertained.
- Your work might have taken you away from your family, a situation that can increase stress and feelings of isolation. You've heard all the clichés before: quality time, getting reconnected, reclaiming your role as a son or daughter. Even though these terms are hackneyed, they express important notions. Just remember that your family values you not because you're a decent cataloger or a topnotch systems analyst, but because you're a member of the clan, you were adorable as a baby, you look like your father, and numerous other reasons that have no connection to librarianship. Your family allows you to be something more than a librarian, which is a relief. In the family setting, you can put that recon project into a better perspective. And you won't worry about it so much.

3.7 If single, accept and enjoy it

- Understand the essential difference between loneliness and solitude. The former involves boring routines and dinner for one in front of the TV. The latter can be a pleasure arising from relaxing afternoons with a stack of the latest fiction, interesting travel and adventures abroad, gourmet cooking, and the freedom to set your own schedule.
- Choose the level of social activity that you're comfortable with. You're not forced to party all weekend or visit your parents every weekday evening. Consider your personal levels of tolerance for different kinds of company.
- You can undergo therapy to deal with shyness, agoraphobia, or stuttering. No one needs to become a hermit because of these conditions.

3.8 Investigate popular stressbusting methods

- Yoga, the Alexander Technique, music therapy, visualization, autogenic training, massage, acupressure, and Qi Gong are effective ways to reduce stress. Of these, yoga is the oldest and most reliable, not only for its relaxing qualities but also for its value as a form of exercise. For librarians with back and neck problems, yoga is especially beneficial.
- Hobbies create eustress. Model railroading, herb gardening, painting, needlework, and musical instrument repair are examples of hobbies that have attracted numerous librarians over the years. The purpose of any hobby is to get the hobbyist's mind off acquisitions policies and fee-based service.
- Religious observance can be a good stress-buster. Churches, temples, mosques, synagogues, and other religious institutions provide opportunities to socialize, to participate in community service, and to think about issues that transcend mundane experience. Be warned, however, that religious institutions often maintain libraries, and you might be invited to sort out the catalog. Refuse politely and offer to bake cookies.

If it is to be effective, your stress-reduction program must be an ongoing process, not just a temporary pastime. In essence, you must change the way you live and work, which in the beginning is a stressful thought. But the benefits of stress reduction will quickly become evident, and the burden of your life will be considerably lighter. It will also give you the strength and peace of mind to handle fatuous in-laws and bachelor uncles.

3.9 Walking: The librarian's choice

Of all the forms of stress-reducing exercise, regular walking is the safest, cheapest, and for many of us in sedentary professions, the most effective. You might not enjoy jogging or aerobics because you find them dull or hard on old injuries. Walking, however, can be entertaining as well as easy to tolerate physically. It can be a good way to control weight. Walking stimulates the immune system, aids digestion, and increases energy.

If you walk regularly, be sure to wear comfortable shoes and clothing suitable to the local weather conditions. For longer walks, wear either hiking boots or jogging shoes with firm arch support. Start slowly at first, and walk brief distances of 1 mile or less. Eventually you can increase your regular distance for additional benefits.

Walking can be aerobic or relaxed. Find the pace that suits your mood. Walk with a friend. Take your dog. Leave your mobile phone at home.

4 The Elvis biography has just left the building, and nobody checked it out: A primer on library theft

Every year, libraries lose increasing amounts of valuable books and other materials to thieves. Here's an overview of theft techniques and the ways that you can reduce losses throughout your system.

You and the circ desk staff were chatting with patrons or working at terminals as I sauntered through the stacks and out the library's rear door. You should keep it locked, but on hot days like this you prop it open to allow the breeze to cool your workplace.

For the record, I've just stolen the latest biography of Margaret Atwood, three magnificently illustrated works on French painting, and popular hardcover fiction by John Updike, Brian Moore, and Toni Morrison. Not a bad haul, although it's not my biggest or most valuable. I've removed veritable mountains of items from libraries over the years, and I've bagged some superb collectibles.

4.1 Spot the crook

Who am I? In fact there is no consistently dependable psychological profile for a book thief. I could be anyone: an elderly woman in a bulky overcoat, a boy with a knapsack and rollerblades, or a young mother pushing a stroller. You might assume that street people pose the biggest risk to your collection, but junkies and drunks are not necessarily the most frequent or successful book thieves. Doctors, lawyers, and business leaders have all been caught in the act, as have clergy of every denomination. Students are notorious for library theft, but certain professors are no better. Even library school faculty members have been suspected of stealing: in one case, an archival studies professor was investigated for looting his university's special collections division. He was never charged, but his teaching contract was not renewed. He found a new job in a library in the United States. You have to admire his initiative.

Before you start moaning about the wickedness of the world, consider yourself and your colleagues. Fascinating new titles arrive in tech services every day: have you ever been tempted to take one home and read it before it is processed for general circulation? Isn't such "borrowing" your right as a hard-working professional? Don't you intend to return that title as soon as you've finished it? Isn't reading new titles that arrive a form of professional development?

In most cases, these questions are attempts to justify theft of public property. You are not entitled to take books from your library without going through the standard check-out procedure. Is there a library that allows staff to do so? Has any board approved such a policy? Is it warm in here, or is that a blush?

4.2 Why steal?

In your defense, there are different motivations for stealing books from libraries, some more pardonable than others. If you've taken home unprocessed titles for a fast read before the public takes its turn and there is no doubt that you will return those titles promptly and in good condition, then your crime hardly merits eternal damnation. If, however, you're selling those titles to second-hand book dealers for extra and untaxed income, you're no better than any other book thief.

Theft for resale is common, despite the library's marks of ownership. It's easy to remove many labels and plastic covers, and even if a book is stamped, it is still saleable. Second-hand book dealers are often willing to buy and sell ex-library materials, since to dedicated readers and collectors a former library copy of a desirable title is better than none. A "withdrawn from collection" stamp reduces the value of a book, but not completely. A first edition of a rare book is attractive, even with labels and stamp impressions. Incidentally, for about $20 many stamp manufacturers will produce an exact copy of any stamp that your library uses. For the best—or worst—use of a bogus stamp, read on.

Intellectual packratting is another motivation for book theft. There is so much to read and understand; there are so many different stories and visions and ways of thinking to investigate. Even if I don't have the time and inclination to study Milton and Dante and William Blake, at least I can have their works on my shelves in preparation for the day when time and inclination coincide and I'm ready to visit their versions of heaven and hell. But books are expensive, and shoplifting from a busy, security-conscious megastore such as Waterstones is risky. So I might as well "borrow" Milton and Co. indefinitely from the library. It's easy to do, and how often do people end up in court for stealing library books? (By the way, has anyone ever served a jail sentence for stealing Milton's *Paradise Lost*?).

Preventive theft occurs most often on university campuses. Jealous Professor Smith wants to stop up-and-coming Professor Jones from finishing a research project, publishing a learned article, and securing another large grant. Smith steals certain titles from the university library. He knows that Jones needs these titles, and now Jones must obtain them through interlibrary loan, which takes time and adds to his expenses. Using a razor or exacto knife, Smith might also remove articles, diagrams, maps, and tables from other titles that he assumes Jones will need. As the scramble for research funding becomes more competitive, academic librarians have good reason to fear the increasing depredations of their collections at the hands of the Smiths.

Finally, theft is fun. It's a game, a challenge, a sport. When I walk out your door with Hemingway's *Death in the Afternoon* under my coat, I can almost hear the crowds shout *Olé*! Maneuvring through your collections, avoiding the gaze of the library assistants, careful not to attract the attention of honest patrons, I feel a rush of excitement. This is my way of winning in a world that made me a loser. And it's so much safer than robbing a shop.

4.3 How it's done

A library thief develops a modus operandi or style of stealing that includes one or more of the following techniques.

4.3.1 Under the coat

A traditional method of concealing items before walking out an exit. "Under the coat" includes hiding items in pockets, tucking them in an armpit, and securing them close to the body under a tightened belt. Very effective in libraries without Tattle-Tape systems. Not advisable during the hot summer months, since persons in overcoats look suspicious indoors during heat waves.

4.3.2 False briefcase compartment

Usually in the form of a false bottom space or a concealed side pocket. False compartments are good for hiding smaller items, but consider other techniques if you plan to rip off large-format volumes.

4.3.3 The chutzpah glide

As its name implies, this technique requires shameless audacity as well as an ability to keep cool and look inconspicuous. You simply walk out the library exit carrying the items that you're stealing. You make no effort to conceal them. You can tuck scraps of paper in those items to look as though you've marked your place in works that you've been studying for an extended period. You seem to be a mature and serious student. Above all, you must appear confident. Your move toward the exit must be calm, unhurried, and nonchalant. You must not attract attention. This technique is effective during busy times when staff are attending to the needs of honest patrons.

4.3.4 The unprotected exit

In libraries where you are threatened by Tattle-Tape or vigilant staff, unprotected exits are ideal for a fast and inconspicuous departure. Look for side and rear exits, basement doors, and any windows large enough to allow undetected egress.

4.3.5 The unprotected and unmonitored staff area

To leave the library by cutting through a staff area requires chutzpah, but you'd be surprised how easy it is to wander unchallenged through the back rooms, basements, and miscellaneous storage areas of many libraries. There is almost always a staff door through which you and your booty can escape.

Some thieves are bold enough to wear a library staff member's identity badge, available from former staff or custom made by the same people who make stamps and other business paraphernalia. Because of high turnover and large on-call staffs in many library systems, it's unlikely that anyone will immediately peg you as a fraud. In fact, well-mannered librarians might offer to direct you to the most convenient exit. After all, good service is the hallmark of the profession.

4.3.6 The baby carrier scam

A variation on the theme of the false briefcase compartment. You don't even need a real baby, as long as you use a carrier that effectively conceals its contents. Traditional prams, strollers, wagons, and snugglies have been employed in successful thefts. The larger the carrier, the more you can carry out, and with a big pram you have room for bulky nonbook items such as computer hardware, A-V equipment, and artworks. It's wise to act like a real parent when you use a baby carrier. Take time to adjust that blanket and coo softly at those darling car manuals and CDs that you're about to rip off. At least you don't have to change their diapers.

4.3.7 The phoney "withdrawn from collection" stamp

First, visit the library sale shelves. For $1, buy a discarded book that contains a clear stamp impression of "withdrawn from collection." Have an exact copy of this stamp made, for approximately $20. Buy the right color of ink (usually blue or red) from a stationer. Return to the library and select the books you want. Take them into a washroom cubicle and, with your phoney stamp, leave clear impressions in the usual places: endpapers, title pages, etc. Then take them to the circ desk and purchase them as discards for $1 each. The circ desk clerk will even deactivate the Tattle-Tape for you, after which you're free to go. In order to leave endpapers and title pages clear, some thieves use erasable ink for their phoney stamp impressions. Variations on this technique include bogus stamps from other libraries and phoney date due stamps. Be sure to adapt this technique into the procedures of the library that you're… patronizing.

4.3.8 Bag and drop

This technique is frequently used in libraries that employ state-of-the-art Tattle-Tape systems, which are difficult to circumvent. All you need is a bag of some sort to hold the items that you want. When your bag is full, drop it out a window, preferably on a soft and inconspicuous spot such as a bush or a pile of garbage in an open dumpster. If you must drop your bag from a height of several storeys, make sure that it's properly padded to absorb the impact.

"Bag and drop" has been used in some of the most impressive library rip-offs. In one notorious case, the thief was spotted dropping his bag by passers-by. Leaving the library, the thief was presented with the bag by a good citizen, who mentioned that it wasn't safe to drop medieval manuscripts from third-floor windows. The thief agreed and thanked the citizen for his trouble. To paraphrase P.T. Barnum, there's a good citizen born every minute.

4.3.9 Scanner avoidance

Limp in on crutches. Select the books you want and conceal them approximately. Then ask the circ desk clerk to let you exit through a side door because you can't negotiate the turnstile, which just happens to employ a Tattle-Tape scanner. Such a nuisance! The clerk will wish you a speedy recovery as you limp out with your haul. Variations on this theme include the use of wheelchairs, walkers, and electric carts. You can even use a cane, as long as you can communicate to the clerk the idea that passing through that turnstile will cause you pain.

4.3.10 Washroom processing

A washroom cubicle gives you the privacy you need to remove Tattle-Tape or to cut out desirable illustrations, photos, and sections of text. Remember that Tattle-Tape is usually hidden down the spine of a book (a "tweezer job") or pasted onto an inside cover. Tattle-Tape is easily removed from the latter location with a small knife. Once you've removed Tattle-Tape from an item, discard the Tattle-Tape in the used towel bin. Don't leave it lying around the cubicle, since bits of tape will alert library staff to your activities. Heaven forbid. And it's not smart to flush a large amount of Tattle-Tape down a toilet, since it may clog a pipe, cause a flood, and alert the library staff, etc.

As for cutting out illustrations with a razor, this is especially common in libraries with collections of older books or high-quality fine arts materials. Do you remember that series of framed prints that you saw at your local antique dealer's? They're very attractive, and on sale for $700. Did you know that the local university library holds a badly mutilated eighteenth-century volume that once contained those prints? You get the picture. And you can get all the pictures you want with an exacto knife.

4.3.11 The visiting dignitary

Thieves have been using false identities since Satan dressed up as a snake. The visiting dignitary technique has proven successful in numerous libraries as well as museums and art galleries. Usually the thief purports to be a professor from a prestigious university. She appears at the library director's office bearing a phoney business card and forged letters of introduction. May she spend a couple of hours in the special collections vault? Of course she's so distinguished and dignified that no one would dream of checking her briefcase for a false compartment.

Months later, the library discovers that several valuable manuscripts are missing. The bogus professor has sold them to a crooked book dealer in the United States or Europe for a substantial sum. Since then she has hit several other libraries, and no one has questioned her letters of introduction or the purpose of her research. Nor has anyone noticed that her briefcase is made from the finest snakeskin.

4.4 In the event of capture...

Sometimes the techniques that I've described don't work, and you nab me. What next? In fact, convictions for theft from libraries are rare. Librarians are often disinclined to go through the process of laying charges, going to court, giving evidence, and undergoing cross-examination in a case involving stolen fiction worth under $1000. Faced with expanding case loads and enormous amounts of paperwork, the police are seldom eager to devote resources to the conviction of a biblioklept. Thus the risk of a heavy fine or a jail sentence is negligible.

Librarians are left to deal with the thieves they catch in the act. They will find the experience frustrating, since there is so little that they can do aside from issuing stern warnings and banning the thief from the library.

Catching a thief can be dangerous. To lay hands on someone who is technically innocent of any crime might be considered a violation of rights. Moreover, some thieves are desperate: they might resort to violence, for which few library staff members are prepared. It's easy to understand why so many librarians connive at the activities of thieves.

4.5 Preventive measures

So what can you do to stop me? There are eight effective measures to discourage theft in any library. Note that you'll never stop me altogether; you'll simply make my activities more difficult, and eventually I'll move on to easier targets. But consider the following preventive measures carefully. Any single measure creates a nuisance for thieves but, when these preventive measures are implemented together, I'm faced with challenges that I'm not prepared to accept.

1. *Tattle-Tape*
   There is strong anecdotal evidence across North America that Tattle-Tape reduces library theft. There is probably no better deterrent, especially against amateurs and impulse kleptos.
2. *Signage*
   Thieves are discouraged by signs bearing messages such as "advanced security system in operation" and "books are public property." "Thieves will be prosecuted" and "Report all suspicious behavior to the circulation desk." Even a professional thief will be a little nervous about that advanced security system. Is there some new security device in the market that he doesn't know about? Has it been installed in this library? Perhaps it's time to find a branch that isn't so high tech…
3. *Warnings to suspected thieves*
   Thieves hate the idea that they're under suspicion. To tell a suspected thief that stealing books is illegal is no violation of one's rights. Other ways of getting across the message:
   - "We keep an eye on people we suspect of stealing, you know."
   - "I'd appreciate it if you didn't carry that bag into the stacks, ma'am."
   - "If you try to pass through the front turnstile without checking out the books under your coat, sir, an alarm will sound. A very loud alarm."
   - "I wouldn't use that rear door as an exit, chum. I haven't fed my Doberman today, and he's waiting on the landing outside."

   If thieves can employ chutzpah, so can librarians. In fact, a touch of wry humor might be more effective that the sternest warning. You don't need a real Doberman to enhance your security.
4. *Patrols*
   An occasional stroll through your library will discourage thieves who depend on a general lack of surveillance, particularly in and around less frequented spaces such as basements, stairwells, and corners of stack areas. Patrolling washrooms is also prudent.
5. *Hardening*
   Don't leave doors and windows open that should be locked. Don't allow the public to wander through staff areas. Bolt all computer equipment to supporting surfaces. Make sure that all staff members realize that good security protects them as well as

the library's collections. Encourage staff to report all suspicious behavior to library management or on-site security personnel without delay.

6. *Security checks for visiting dignitaries*

He claims to be a professor from Yale. Before you allow him into your special collections vault, check with Yale to see if he's the real thing. One telephone call or e-mail inquiry might save you much grief. Let that professor know that you are obliged to check his identification. If he's a fraud, chances are that he won't stick around for long.

If he's impersonating a real professor and he presents top-quality phoney letters of introduction, you can still catch him. One famous thief was caught when a librarian phoned what was assumed to be his residence in Cambridge, Mass. "You say that you're Professor Smith from Harvard, and you want to examine early maps of Quebec," said the librarian. "That's right," said the fraud, with admirable self-possession. "I've already given you my card and letters, and I need to see those maps for the purposes of my historical research on the influence of the Catholic Church." The librarian remained poker faced. "Is that so," she said. "I just spoke to your wife in Cambridge. She tells me that you died 3 years ago. Isn't it a little late to take an interest in religion?" The fraud left the library in unprofessorial haste.

7. *Informing other branches and libraries in neighboring regions*

If you suspect that a thief is active in your library, let your local colleagues know. If possible, describe for them the thief's appearance and techniques. Some libraries circulate written descriptions of thieves. When a thief realizes that she is recognized throughout your system, she will move on.

8. *Informing local book dealers*

Share with local book dealers the same information that you give your library colleagues. If dealers refuse to buy suspicious items that might have been stolen from your library, a thief's income will decrease. Dealers are usually happy to cooperate with libraries in frustrating thieves and are grateful when librarians protect them from receiving stolen goods.

4.6 On the move and on the make

So many books, so little time. I don't mind telling you how to stop me because I know that in many libraries apathy rules. I've robbed lots of branches that don't even turn on their Tattle-Tape scanners. Librarians write memos about me, and their managers don't even bother to read them. Doors and windows remain wide open; I come and go as I please.

And business is good. Tomorrow I'll hit two public library branches in Victoria and three in Vancouver. Then I'll visit the universities of Saskatchewan and Manitoba, which have some lovely historical material on the shelves. (Not for much longer.) Of course the University of Toronto and TPL are always worth a visit.

Then on to Halifax: no day is complete without a little maritime larceny. You might call it a cross-country check-out, except I have no intention of checking out a single item.

Snakeskin is popular these days. Wouldn't you agree?

The data backup process: An overview for saboteurs

5

The author is pleased to present an excerpt from the recently published third edition of Fundamentals of Library Sabotage, *a popular textbook for disgruntled catalogers, irked researchers, angry patrons, and LIS students looking for practical ways to realize their end-of-term fantasies.*

If all libraries backed up their electronic data, the saboteur's job would be more difficult, and the results less satisfactory. Fortunately, many libraries either neglect to back up the data for their electronic catalog and administrative operations, or depend on a backup process that is ineffectual. Accordingly the saboteur is in an excellent position to deprive numerous libraries of their data, and to disrupt library service worldwide.

Library administrators recognize the value of their data, and allocate substantial funding for the development and maintenance of automated systems. Rarely, however, do they consider the risks that prevail in all high-tech environments. Natural disasters such as high winds and storms are constant threats, as are fire and flooding. But human folly can be an even greater danger to computer systems. Librarians wreak havoc through operator error, ignorance, or sheer clumsiness. They trip over wires and accidentally pull servers off shelves. They leave hard drives to bake in the sunshine that pours through their office windows. Unintentionally they introduce the latest viruses into their networks, and open the most suspicious-looking e-mail. In such circumstances, the saboteur has countless unwitting allies. It is reasonable for him to assume that, if he does not destroy a library's critical data, someone else will.

In light of all prevailing risks including sabotage, the survival of data depends on the process by which they are backed up—i.e., duplicated for the purposes of recovery.

In the absence of a backup process, the saboteur can destroy data by a variety of methods, including straightforward erasure while nobody is looking, causing a power outage (the fuse box is in the basement), planting a virus or worm, or simply stealing the server, which is on a shelf in that stifling little room beside the broom closet. Once unbacked-up data is erased, it is gone forever. If the saboteur is an insider, she can stand back and savor the fruits of her labor: the frenzied staff meetings, the hollow excuses, the director on stress leave, the patrons who want to know why the system is down again, and why the library ever got rid of that good old card catalog, and the trustees who get sulky when they learn that the records for acquisitions and fines have disappeared. If the saboteur is an outsider, he can read all about it in the local newspaper: LIBRARIANS' WATERLOO: DATA DOWN DRAIN.

The challenge to the saboteur from an ineffectual backup process can be negligible. To wipe out critical data, the saboteur can exploit any of the following weaknesses:

- The backup process is described in the library IT systems manual, which nobody reads.

If the IT systems manual contains the only description of the library's backup process, and nobody reads the manual, then the chances are that the backup process has not been implemented. Saboteurs should be on the lookout for manuals in three-ring binders that remain wrapped in plastic. Another indication that a manual has been ignored is its combination of thickness, jargon, and badly rendered diagrams. The resulting lack of clarity and concision will ward off all but the most dedicated systems technician.

- Critical data are backed up every now and then. Maybe.

Effective backup requires a regular schedule and a log. To determine the regularity of a library's backup schedule, the saboteur should listen for remarks such as:

"In this library we back up the data now and then. I think. At least we used to. Well, we used to back up the catalogue in the last library where I worked. I think we did. Or did we? Yes. I'm sure. I still have the floppy in my glove compartment. Or do I? Sorry, that's an electronic copy of my husband's dissertation… Where was I?"

An irregular backup process creates ample opportunities for the saboteur, who can wipe out data and rest assured that the backups will be incomplete.

- Critical data are backed up annually.

Annual backups quickly become out-of-date. If sabotage forces the library to use its antique backups, librarians will be pushed through a time warp. Note that daily backup places no burden on most systems departments, and puts the saboteur at a serious disadvantage.

- Inessential or worthless data are backed up, while critical data are not.

While systems librarians back up data regarding football pools, the director's allergies, and recipes for potato salad, the saboteur can destroy the electronic catalogue without fear of its eventual recovery. Librarians have also been known to back up old backups rather than current electronic resources. The saboteur can ignore those old backed-up backups, and concentrate on fresher targets. As for the unbacked-up backed-up backups…

- Backup media are stored on site, in some cases in the same work area as the original data.

Here is a superb opportunity for the saboteur. He can destroy original data and their backups in one fell swoop, without leaving the scene and having to struggle with off-site vaults and security staff. Note that many security planners do not consider data properly backed up unless backup media are stored off-site.

- Backup media are held in low-quality off-site storage areas.

For the saboteur's purposes, the best off-site storage areas for backup media include glove compartments, the assistant director's briefcase, the chest of drawers in the bedroom of the programmer's summer cottage, and the skid row warehouse where pigeons nest on the rafters and a fresh breeze blows through the broken windows. Secure off-site data storage is a severe impediment to successful sabotage.

- Backup media are untested.

Has anybody checked to make sure that backups actually contain all the data that they should? Are there any glitches? Has anybody run a virus check on the backups? The saboteur should be eager to find out.

- Backup media are obsolete.

These critical data are backed up on a floppy with the circumference of a Frisbee. But wait. These data are backed up on cards with holes in them and these data are backed up on clay tablets. What are the chances of recovering systems operations using these media? The older the backup medium, the less a saboteur needs to worry about it.

- Backup media are lost or mislabeled.

If the librarian cannot find the backups, the library's data are not backed up. If the label says ELECTRONIC CATALOG BACKUP TAPE and mentions the current month, the librarian expects catalog backup data and not a list of Adult Services subcommitees from 1975. The saboteur can "lose" the library's backups, but it is even simpler to mislabel backups and send desperate systems librarians on a fruitless search. Some of the most effective mislabeling involves nothing more than an instruction such as DO NOT ERASE UNTIL 1989, or DUPLICATE STATS FOR XMAS PROGRAMS—1981–1985. The saboteur is encouraged to be creative.

- Responsibility for the production, quality control, and auditing of backups is not clear.

Who is in charge of the library's backup program? Who makes sure that backup media are shipped off-site at the end of the workday? Who makes sure that the process is smooth and reliable? If the answers to these questions are vague, the saboteur should look for gaps in the backup process. Informal arrangements are acceptable for some procedures, but not for the protection of critical data.

The saboteur must understand the implications of a backup process that is truly effective. Happily these are scarce, but they could become more common as libraries strive to enhance their security measures. Should the saboteur be faced with a top-notch data backup process, he must accept the possibility that the library's critical data are beyond his reach. He is advised to consider the alternative methods of sabotage that are discussed elsewhere in this book, particularly in the earlier chapters on arson, sprinkler sabotage, and website hacking. There is always a way to disrupt library operations, but the saboteur's success often depends on equal amounts of skill and patience.

5.1 Saboteur's bane: High-quality off-site storage

Top-notch off-site storage facilities can be serious impediments to the saboteur. Those intent on destroying library backup media held off-site could be thwarted by a storage facility with the following features:

- *Flexible media accommodation*: The facility can receive, store, and conserve the volume and kinds of media that the library uses for backups.

- *Acceptable design standards*: The facility meets design standards that have been recommended by trade and professional associations such as ARMA and ACRC. The facility conforms to all local building, security, and fire codes. In British Columbia and parts of Quebec, the structure of the facility can withstand earthquake loading.
- *Secure transportation*: Backup media are transported to and from the facility in unmarked vehicles. Drivers have been instructed not to allow unauthorized personnel to ride in the vehicle or inspect its contents. Drivers will not answer questions about their schedules or destinations.
- *Trained personnel*: All facility personnel are trained in the handling of backup media.
- *Security measures*: All facility personnel have passed a security check. The facility has access controls including alarms and electronic monitoring. Access is restricted to facility personnel; library clients can inspect their media in a visitor's area.
- *Access provision*: The facility can provide 24-h access to backup media in case of emergency.
- *Strategic alliance*: The facility plays an essential role in the library's security and disaster recovery plans.

Saboteurs should remember that significant disruption of library operations depends on the destruction of both the original electronic media and their backups. If backups are allowed to survive, the saboteur will ultimately fail.

What's a little larceny between friends? Confessions of an info-thief

6

Your library's borrower records are worth a bundle. Are they secure? The following is a chapter of an autobiography that is sure to attract a large readership, particularly among law enforcement officials.

Midway through my career, broke and desperate, I was obliged to find new ways to earn a living. My library wages were low, so I decided to go prospecting for the electronic gold in borrower databases. Simply put, I became an info-thief.

I specialize in the acquisition, interpretation, repackaging, and distribution of the personal data that borrowers give public and academic libraries. The scam is straightforward. To become a borrower at your library, a person surrenders information about himself and receives a library card. His name, address, postal code, and home telephone number constitute his basic entry in your borrower database. Sometimes libraries include other data—what I call the "delicious extras"—such as the borrower's age, social insurance number, business address and telephone numbers, e-mail address, and information concerning spouse and children.

By themselves, these data are saleable for the compilation of mailing lists. But the most valuable data are those related to borrowing tastes and practices. Consider the record of Jane Doe. She lives in a polite suburb of Winnipeg. From her address and postal code, I can determine that she occupies a house; other data indicate that she is a mother (probably single) of Brad and Bart, ages 5 and 9. She works for the Government of Manitoba as an assistant manager in a ministerial systems department. But here's the gold: in the past 3 months, Jane has borrowed or reserved 9 books on buying and selling real estate, 3 on used car maintenance, 6 on cheap Mexican holidays, 3 on mutual funds, and 2 thick volumes on caring for elderly and disabled parents.

On the Borrower Application Form, Jane has noted that she is 45 years old. That age suggests one reason for her borrowing various health and fitness manuals as well as a stack of workout videos. Jane is careful about her diet, too. Last month she borrowed a dozen books on low-fat cooking, cholesterol risk, and salad construction. Jane has also borrowed several mysteries, the latest Atwood, and George Eliot's *Middlemarch*. Good for her. She's free to pile high her bedside table with fiction while I get down to the task of interpreting her data.

First, her interest in real estate books would attract the attention of any local realtor. She might want to sell her current house and move to a big condo with Brad and Bart. She might also want to take them with her on that cheap trip to Mexico, and there are numerous travel agents in Winnipeg who'd like to help her with the arrangements.

Robertson on Library Security and Disaster Planning

Meanwhile, the brokers of mutual funds can't wait to talk to her about the importance of getting into the market as soon as she can.

And then there are Jane's mom and dad. It may be time for them to consider renting a suite in that splendid seniors' care facility in Ottawa or Montreal, or perhaps Flin Flon. Or perhaps they'd like to move to a private hospital in Toronto or a "Snowbird" compound in Florida. The managers of these facilities would just love to talk with Jane about her parents and what they can do for them.

Young Brad and Bart will interest a plethora of suppliers. Whatever Jane buys for her sons has its manufacturer and distributor. They'll be interested in Jane's overall customer profile, which they can derive from her address, age, marital status, and the gender and ages of her offspring.

Finally, there's the fat factor. Across the continent there are distributors of books, videos, software, and exercise equipment that Jane might like to consider.

Remember that I'll steal your entire borrower database, not just one borrower's records. There are hundreds of Janes in the database of any large public or academic library in North America. These databases provide a first-rate demographic picture of a community, its residents, and their interests. Once I've interpreted the data and determined who will want to buy what, I can repackage names, addresses, and other details for distribution. Realtors like lists of personal information organized by neighborhood. While they will accept paper lists, they prefer a data stick or CD. They're always interested in fresh leads, and as long as I look honest they won't ask questions about the source of my information. In any case, I can always string them along and tell them that as an information professional, I have numerous resources.

I may charge $5000 for information related to a single neighborhood—or even a single street. A realtor need only sell a couple of small properties to justify the cost. In cities such as Toronto and Vancouver, where real estate prices are exorbitant, I can double or triple my price. And I can resell my data many times, to realtors working for different companies. At least one of them will telephone Jane and offer to help her sell that old house and buy something new and Brad-proof.

And what about Jane's used car? She wants to maintain it, but perhaps she might consider a splendid preowned item with umpteen options and great financing. It's waiting for her at a lot where the manager has paid me $2500 for a list that includes not only Jane's personal information but also the personal information of hundreds of other Winnipeg residents (i.e., library patrons) who drive clunkers.

The travel agent has just paid me $3000 in small denominations for a list of Winnipeg residents who have recently borrowed tourism titles. And I've just sold a dozen copies of a list of library patrons who have borrowed titles on stocks, bonds, and the joys of a bull market to a regiment of financial advisers. In fact, I have my doubts about the market, since those advisers wouldn't pay me any more than $1000 per list. But several of them told me that they'd buy an updated list in a couple of months, so I won't complain.

Jane will wonder how so many health product distributors found out about her. Every day she receives a heap of junk mail advertising everything from organic seaweed to exercycles. A woman telephones her with a pitch for bee pollen wrinkle control cream. A man sends her an e-mail inviting her to a seminar during which she can rediscover her inner child.

Meanwhile, my financial health continues to improve, since the trendy fitness mob can't survive without continual marketing and promotion, and lists such as those that I steal from libraries are invaluable. Generally, I can charge between $1500 and $3000 per list, depending on the size of the list.

Most librarians would be shocked at how easy it is to steal their borrower data. Often all I need to do is ask where the appropriate server is stored, and my beloved colleagues will tell me. Data security is dangerously unsophisticated in most libraries, and many systems areas are unlocked. Or the operations room is locked, but the storage area for the tapes and other electronic media is in a closet that the janitor leaves open because it's a nuisance to fiddle with the key, and besides, who would want to steal any of that stuff anyway?

My latest approach is to ask a librarian how I can be sure that my personal data are secure. She will tell me that I can relax, since all of that information is kept on the fifth floor in a very safe place. It's backed up, too. And the back-ups are in a file cabinet beside the Technical Services Department. Look, the cabinet is even labeled "back-up media." And every loan is recorded, I can be sure of that. I'm very grateful for this information. Now, when do the tech services staff leave the office for lunch?

These days I have little trouble obtaining the data I need. The future looks bright, although I must admit that recent privacy legislations might cause problems. For example, if librarians were to review current government guidelines and recommended measures to protect personal information, they might start safeguarding their borrower data more effectively. Systems librarians might ensure that all back-up media were stored in secure locations and strangers were not permitted to enter any systems area. Internet access to borrower records would be tightly controlled so that hackers could not break into databases. Worst of all, librarians wouldn't talk to members of the public about their data management practices. Those loose lips have kept me in business!

If librarians do improve their data security, I could end up looking for another racket. Perhaps I'll retire and join Jane in Mexico. After all, her travel agent has been one of my more enthusiastic customers.

7 Clearing the air: Improving indoor air quality in libraries

Poor air quality is a problem in many libraries. Causes vary from building to building, but often the source of bad air is mechanical. Heating, ventilation, and air conditioning (HVAC) systems can become inefficient over time, and more expensive to keep in good condition. In an age of budget cuts, libraries sometimes permit less frequent maintenance of furnaces. Old filters deteriorate and let larger amounts of dust and other particles circulate throughout buildings. Dust is to be expected, but not in quantities that remind staff and patrons of construction sites.

Aging HVAC systems have achieved notoriety. The furnaces in various Carnegie buildings provide great blasts of heat only on hot summer days; in December they shut down to give staff members an opportunity to show off the latest in arctic survival gear. Despite sophisticated controls and monitoring, some recently installed systems are equally inefficient. Architects, engineers, and property managers suggest that sometimes new HVAC systems need a few months to "fit" a building: in other words, those responsible for the HVAC need time to learn how it works in a particular structure with specific uses and schedules. Adjustments might be necessary before temperature and air purity attain the desired levels.

7.1 Bacteria, etc.

Since Legionnaires' disease hit the headlines in the mid-1970s, librarians have been increasingly concerned about the spread of harmful bacteria in their workplace. The possibility of an influenza pandemic, such as that which struck in 1918–1919, contributes to the fear of illness contracted on the job. While HVAC systems can be implicated in the transmission of disease, they are not the only culprits. HVAC strongly influences air quality, but does not control it completely.

Patrons bring more to a library than reference questions and the desire for good fiction. They carry colds and flu. On a single shift, a children's librarian can be exposed to dozens of careless coughs and sneezes. It takes strong rearing to teach a child to cover his or her mouth before sneezing, and some people never learn. Reference and circulation desk staff can be equally vulnerable to sick patrons.

While many librarians take time off to recover from the colds that they contract, others insist on showing up to work. The latter contribute to poor indoor air quality and the continuation of the flu season at their workplaces. At present, in many libraries, staff members who refuse to stay home with their infectious illnesses are praised for their dedication. Eventually, however, their managers might decide to protect uninfected staff by insisting that cold and flu sufferers take time off. Ridding the workplace of airborne pathogens is an obvious way to reduce absenteeism.

Robertson on Library Security and Disaster Planning

Some patrons wear perfumes and lotions that librarians find offensive. The recent popularity of "attack perfumes" has given rise to a number of complaints from staff at libraries, especially in downtown cores. The odor of these perfumes can linger in stack aisles and reference areas for hours after their wearers have departed. Some librarians consider the odor nauseous. Unsure how to deal with it, they will often wait for it to dissipate rather than lodge a complaint or call for assistance from property managers.

Not all scents are intended to please. Libraries occasionally attract patrons whose inadequate hygiene results in unpleasant body odor. Most librarians are unsure of how to deal with a foul-smelling patron, and will not act until other patrons complain. Asking a person to leave the library because she does not bathe regularly is an unpleasant task. That person might take offense, and a librarian's sympathy for homeless or otherwise unfortunate members of society makes the situation even more difficult.

7.2 Reactions

The effects of bad air quality in libraries can be divided into three broad categories. First, there are mild reactions, including red and tearing eyes, dry skin on scalp, face and hands, and fatigue or sleepiness. These reactions can become chronic, and are exacerbated by outdoor pollutants and stress. Librarians will often accept such minor irritants as "part of the job."

Second, there are allergic reactions. These include rhinitis that leads to a runny nose and sinus congestion, along with more advanced irritation of the eyes. More severe skin rashes similar to eczema are also associated with allergic reactions. Sufferers note that their symptoms become less severe when they leave their workplace; when they go on holiday, their symptoms can disappear altogether. Bad air quality is frequently identified as a problem in the library by staff members who, on returning to work after a holiday, experience a recurrence of allergic reactions.

Third, there are reactions that resemble and feel like influenza: fever, aching muscles, and chills. In many cases, these reactions are triggered by airborne microorganisms that ventilation filters fail to catch. Toxic mold is a potential cause of flu-like reactions, and has been implicated in more serious cases of respiratory infection.

Any reaction to bad air can include headaches, throat irritation, and asthmatic symptoms. Numerous reports in medical literature and popular media indicate an increase in asthma among adults, especially those who work in dusty environments. Any of these reactions can lead to decreased concentration and substandard performance of duties. After all, it is hard to read fine print and answer reference questions when one's eyes will not stop watering and one's headache keeps coming back.

7.3 Improvements

Improving a library's indoor air quality involves several steps. The first is the investigation of the HVAC system. Technical specialists should examine all HVAC components and their documentation, particularly service and maintenance records. Often an investigation will uncover weaknesses such as worn-out parts, faulty installation, and clogged filters. Sometimes the system needs simple repairs and replacement of parts, but there is always the chance that a new HVAC system is necessary. Administrators fear the expense, but can justify it on the basis of occupational health and safety requirements. The alternative is to tolerate deteriorating air quality, increasing absenteeism, declining job satisfaction, and lower productivity.

The second step is to assess other potential causes of poor air quality, including toxic mold, inadequate janitorial service, unacceptable levels of sanitation, and dust build-ups in storage areas. It is important to acknowledge all possible contributors to air quality problems, and not to underestimate them. Remember that unhealthy airborne agents are often invisible to the naked eye.

Interviewing staff members about their air quality concerns is the next step. Often they will make useful observations on the air quality in their workspaces, and suggest ways to improve it. They might also complain of illnesses arising from bad air. Objectivity demands that we try to separate those symptoms that are undoubtedly connected to high levels of dust from symptoms that are more likely due to the sufferer's heavy smoking habit or lack of exercise. It is difficult to determine with absolute certainty the cause of any symptom, and in a number of cases we will be forced to rely on a staff member's assurance that a particular HVAC deficiency has resulted in her headaches and sore throat.

7.4 Welcome to the Caribbean

Having gleaned the necessary information from staff members, the library can develop a strategy to improve its indoor air quality. Substantial changes might be required, and administrators should consider ways to justify them. (In one case in the United States, the director of a public library threatened to sue city hall unless a new HVAC system was installed in her library. She complained of increasing asthmatic symptoms, and promised to tell her story to the media. City hall quietly replaced the HVAC system and sent the director on a holiday to the Caribbean. She returned in good health, and remains free of asthma.)

Other parts of the strategy might include enhancement of janitorial service and more frequent cleaning of rugs, moisture remediation to destroy mold, and removal of dust-gathering waste such as old papers, boxes, and furniture. To increase air circulation in specific workspaces, small fans can be installed wherever staff members need them.

As for malodorous patrons, they must always be treated courteously, but the library can set firm policies regarding body odor and heavy perfume. Following the lead of hospitals, some libraries post signs announcing "perfume-free zones." If patrons bristle at the suggestion that their expensive scent is unacceptable, librarians can point out that not even the director is permitted to wear perfume in the library.

Diplomacy is essential when dealing with patrons whose body odor is noisome. Some librarians give them the address of a nearby community center or YMCA where they can shower cheaply or for free. In other cases, librarians have called for assistance from local social workers or shelters for the homeless, who are familiar with the patron and his or her circumstances.

Libraries are among the most heavily used facilities in the world. In workspaces that are open to the public and in constant need of mechanical maintenance and cleaning, air quality problems are bound to occur. There are effective solutions, but they must be renewed and adapted to meet new challenges. They don't give those perfumes brand names such as "Poison" for nothing.

Downsizing the binder: Effective security planning for libraries

8

Summer, 2003: As I write, thousands of librarians in Ontario and the northern United States are struggling to provide service during the largest power failure in history. Meanwhile, on the other side of the continent, huge forest fires in British Columbia are damaging town sites and leaving entire communities homeless or displaced. Via e-mail, a librarian in the southern BC city of Kelowna has informed her colleagues about smoke damage from the enormous blaze that lit up the sky on the city's outskirts. She asks about security procedures following an evacuation. How could her library ward off looters? Is there any way to protect the library's assets?

No matter where you work, security problems are bound to arise. They might be small and inconsequential, like graffiti in the washroom. Or they could be serious: a virus that destroys the OPAC, a violent patron, a stalker in the parking lot. Library administrators deal—or fail to deal—with these problems in different ways. They can ignore them. They can ask somebody else to handle matters, such as the facilities manager or the police. Or they can set up a security committee that will meet, discuss the situation, produce minutes, and make recommendations.

Often the security committee will draft the library's security policy and procedures. Here's where things get complicated. These policies are often based on an ill-defined notion of what constitutes a security risk. Instead of concentrating on criminal or uncivil activities, the committee considers all possible risks, including hurricanes, earthquakes, and other natural phenomena that should be addressed in a disaster plan.

8.1 A comforting—and false—sense of security

The result, which might take years to develop, is a mishmash called the Library Security Manual. In my experience, this monster takes the form of an enormous three-ring binder. Occasionally it appears online, where it is rendered toothless by any power outage or virus that circumvents your firewalls. Most often, the LSM holes up in a drawer in the staff room. No one pays any attention to it until something goes wrong, at which point library employees spend time hunting it down and leafing through it, only to discover that it is out of date, incomplete, or irrelevant.

Nevertheless, the LSM has served a definite purpose. It has given library employees a comforting if false sense of security. As long as a security problem does not become life-threatening, and employees use common sense to deal with minor matters, the LSM will do no harm. But if a problem is potentially dangerous, that binder can become an impediment to appropriate action.

For example, let's assume that an obstreperous patron has arrived at the reference desk at your library. He is drunk. He utters threats and appears to be on the verge of attacking the librarian. In this instance, the LSM binder is close at hand. As the patron reaches for the knife in his pocket, the librarian opens the binder. She notes that it begins with a section on sprinkler leaks. Then it describes what to do in an earthquake. After that is an inventory of washroom keys, a bibliography on book theft, and a list of emergency telephone numbers last updated in 1992.

If this incident occurs anywhere in Canada except the seismic zones of British Columbia and parts of Quebec, why does the librarian need earthquake instructions? She doesn't, but because much of the LSM has been copied from one produced for a Vancouver library, she now knows that it is unwise to run in or out of a building during a quake.

Our drunken patron has been waiting patiently, knife in hand, for the librarian to finish reading about the correct response to an earthquake. Does this scenario make sense? Is it reasonable to believe that the LSM would be useful in these circumstances, or that a drunk with a knife is willing to wait for the librarian to complete her review of the LSM?

8.2 Fostering a security culture

The lesson here is that you can't rely on your LSM alone to keep you safe, especially if its security information is mixed in with disaster preparedness pointers, first aid tips, instructions for using a fire extinguisher, and feel-good policy statements that "put employee safety first!" What's even more important is a heightened sense of awareness of security needs and procedures in your library. As more employees attain this awareness, the library will develop a security culture that renders fat binders obsolete.

What is the most effective process for developing a security culture? Start with an assessment of actual security risks in your library. Remember that you're focusing on criminal and uncivil activities, and their effects on library employees and operations. In a North American public library, your assessment might include:

- Theft of personal property: purses, briefcases, laptops, overcoats
- Theft of books and other library materials
- Mutilation of library materials
- Vandalism, including tagging
- Breaking into restricted areas
- Eating and drinking violations
- Inappropriate behavior toward library employees and other patrons
- Inappropriate behavior in washrooms
- Flashing
- Uttering threats
- Carrying concealed weapons (e.g. knives)
- Stalking patrons and employees
- Distribution or use of banned substances
- Hiding in the library after closing time
- Hacking into library systems
- Robbery or threatening behavior in parking areas

8.3 Assessing and mitigating the risks

The frequency of specific kinds of incidents varies from library to library. If you're close to a neighborhood where drug dealers conduct business, expect them to investigate your library for sales opportunities. When a fast food outlet opens down the street, you might find more burger wrappers and pop cans in your stacks. If the bar on your block offers irresistible happy hour deals, prepare to deal with the excessively happy.

You might not need to worry about these particular risks, but no library is absolutely secure. Inevitably there are exposures, particularly in our wired age where even small libraries have websites. Data tampering and loss is a major concern for all libraries. Fortunately data backup is a simple and easily managed process that any automated facility can implement.

Data backup is a form of mitigation. For each security risk, there will be mitigation measures. It may be difficult to eradicate a risk, but it is usually possible to mitigate it by instituting new procedures, installing security technology, or adopting a combination of these.

After your security assessment is complete and before you begin a mitigation program, you might wish to develop a library security policy. Keep it simple. It needn't be more than a page. In it, you can state your library's definition of security and your intention to maintain high security standards for all employees, patrons, assets, and facilities. Remember to make a distinction between security matters and occupational health and safety audits, disaster planning and crisis management. The latter are essential, but to mix them and security together is to muddy your library's overall risk management.

Your mitigation program should address every security risk that you've identified in your assessment. Some risks have immediate mitigatory solutions. For example, you can decrease loitering in your washrooms and parking areas by setting up patrols and installing better lighting. Regular updating of firewalls and antivirus software is essential for systems security. Locking off restricted areas will discourage most intruders. Other risks are more difficult to mitigate. Tattle-Tape systems don't always discourage book thieves. Many have learned how to remove the tape from book spines with tweezers. Others leave the library via staff doors and windows. Nevertheless, libraries can ban books and other materials from washrooms, and thus frustrate thieves who need a cubicle in which to remove that spine tape. Employees can be put on alert for known thieves, who do not like attention and will eventually avoid libraries that recognize them for what they are. These solutions are not perfect, but they are the first steps to manage risks and control losses.

8.4 Keeping up with changing needs

Often it is not possible to implement a mitigation program all at once. Developing a security culture takes time and money, which are not always instantly available. Thus it is often necessary to spread the implementation of new security technology and training programs over a couple of years or more. Too many libraries have treated security like a one-time purchase, which it is not.

Rather, it is an ongoing process. The security procedures that you adopt now will not be effective indefinitely. With the development of new library facilities, services, and technology, you will be forced to respond to different circumstances. Security risks at your sites might change, becoming greater or less prevalent. You must deal with your current circumstances whatever they include, and so regular updating of security procedures is necessary.

Updating is a task often left to consultants. In my consulting practice, the goal of auditing and updating a library's security plan is to reduce its bulk. I have no use for enormous binders that nobody wants to read, and take pleasure in reducing vast heaps of verbiage to a slim file of point-form instructions. Even these are no replacement for intelligent action and the resolution of security problems in the library's best interest.

And what about that risk of library looting in Kelowna? Highly unlikely. Looters prefer retail outlets. Besides, even Kelowna's criminals will obey an evacuation order, especially if the blaze gets close.

Risks, rules, and enforcement: Enhancing child safety in the library

9

Nothing fascinates 4-year-old Tanya more than the electric front door of the Vancouver Public Library branch that she and her mother visit on their way to the supermarket. Tanya presses the big red button, and magically the door hums open. Her mother is pleased that Tanya responds so well to the opportunities in her environment. Meanwhile, the branch staffers worry that Tanya's fingers might get crushed as the door closes. A library assistant has told Tanya's mother that the electric door is for the convenience of patrons in wheelchairs and strollers, and that little fingers could come to grief on the door's edges. But Tanya's mother says that the library assistant is being "bureaucratic" and that her daughter's high intelligence eliminates any risk of injury at the library. Besides, Tanya's parents have paid for the library with their taxes and should be free to use it as they wish, without petty harassment.

Apparently Tanya did not inherit her intelligence from her mother, whose reasoning is weak. But according to the library staff, there's no point in insisting that Tanya refrain from pushing that button, since she and her mother will ignore such demands.

9.1 Safer than most places

Public libraries are much safer than most other high-traffic facilities. Particularly for children, libraries contain fewer threats than malls, parks, and beaches. The roadway outside the library is far more dangerous to children—and everyone else—than the reference room, stacks, and storytime area. We tend, however, to exaggerate the likelihood of certain risks and ignore others. For example, librarians fear sexual predators who prowl through children's departments in search of victims. According to popular wisdom, these monsters loiter around washrooms or in secluded stack areas, waiting for the young and innocent to fall into their clutches.

Certainly predators exist, and many of them rely on libraries, but rarely for their prey. Libraries are too busy for a predator's comfort. One never knows when a patron or a librarian will appear and make trouble. Washrooms are potential hunting grounds, but these days they can be as crowded as the checkout queue. Predators usually try to avoid detection, and shy away from confrontation. They may visit the library, but they're more inclined to find victims in other places. Some have come to depend on the Internet for contacts and entertainment; others remain on the edge of the schoolyard or near the play area of a local park.

9.2 "We all fall down"

Worrying about the monsters in our midst distracts us from the real risks for young patrons. Running down a Toronto Public Library aisle, 7-year-old Brian trips, falls, and gashes his lip on the side of a carrel. Eleven-year-old Stephanie trips over her new shoes and falls down a flight of Vancouver Public Library steps. Meanwhile Ryan, aged 9, scales the heights of fiction by crawling up the side of a shelf in Winnipeg. Alas, his reach exceeds his grasp. He hits the floor and scrapes his knee. His father complains that the library's shelves are unsafe for climbing. The librarian bites her tongue.

Propelling themselves in all directions at high speed, children fall frequently in libraries. Outside on the public concourse or in the parking lot, their older siblings perform astonishing stunts on skateboards. In some cases, libraries could be held liable for any injuries that children sustain while toppling at their sites. Commonly called "slip-and-falls," these accidents usually lead to nothing more than minor injuries. But they can disrupt library activities and raise the noise level, which is already higher than many patrons would like. And at some point a slip-and-fall could lead to a lawsuit that might force onerous changes to the ways that we manage public areas.

Still, we focus on the predators swarming around the felt board and puppet theater. What we need is a balanced perspective on child safety in the library, with an appropriate amount of attention to real risks.

9.3 Assessing the risks

The risks that prevail at your library affect everyone on site, including young patrons. It's wise to consider the different ways in which these risks could threaten children, their parents, and other adults. In emergencies such as fires, bomb threats, power outages, and severe weather, most people will cooperate with library staff members to take appropriate action: to evacuate the premises, not to loiter in parking lots, to assist injured persons as required. But what about children unaccompanied by parents, teachers, or caregivers? Are librarians responsible for them if they refuse to follow instructions?

Unfortunately, some parents put their children at risk by failing to supervise them properly. They will allow even the youngest child to roam freely through the library and to treat the furniture as if it were gymnasium equipment. If a child has difficulty negotiating an escalator, elevator, or turnstile, the parent will sometimes be inclined to hold the library responsible.

Some adults use libraries as babysitting facilities. They tell their child to look at some books or attend a storytime session, then disappear for a couple of hours. Eventually the child might become frightened or feel abandoned. Librarians at the reference desk are familiar with the boy or girl who wants to know where Daddy is, because he said he'd be back soon but he's been gone a long time. Dealing with such situations under normal circumstances can be frustrating, but during an evacuation or emergency closure, an abandoned child can be an unnerving responsibility.

9.4 A Code of Patron Conduct

To enhance child safety, librarians should develop a Code of Patron Conduct that demands good behavior not only from children, but also from their parents and caregivers. In clear and concise language, this Code should require all patrons to avoid bad behavior: running, shouting, skating and skateboarding, vandalism, spitting, and rudeness to staff members. The Code should also forbid parents and caregivers from using the library as a child-storage depot.

To inform patrons of the Code, libraries should post signs to discourage those activities that are common at specific sites. Some branch entrances are particularly inviting to young sprinters, and skateboarders love wheelchair ramps, concrete ledges, and parking lots. Signage in these areas would be intended not only for children, but also for their parents. At some libraries, signage must be brutally conspicuous, but respect for the Code will compensate for a few large signs.

9.5 The librarian as enforcer

Since the mid-1960s, librarians have noticed a decline in patrons' compliance with library regulations. The bodies of literature on library security and on the treatment of problematic patrons have grown in parallel. Questions regarding child safety are a natural outgrowth of the current library environment, and codes of library security, safety, patron conduct, and facilities use have become much more common. Nevertheless, the key to improving all aspects of safety is the librarian.

Fifty years ago, the librarian ruled the library without question. The stereotypical librarian held a vertical finger to her lips and told the world to maintain respectful silence. In fact, the librarian used more than that finger to enforce the rules. She could suspend borrowing privileges, demand that unruly patrons leave the premises, report miscreants to their parents, school principals or the police, even confiscate property such as knives, food, and makeup. There were few concerns about child safety, since young patrons could rarely put themselves in jeopardy while using the library. The opportunities for havoc were almost nonexistent.

In these more liberal times, librarians tend to tolerate bad behavior and to treat obstreperous patrons with what some might consider misplaced respect. We don't want to reprove a child for climbing on furniture or practicing sprints in the stacks, and we don't feel comfortable in rebuking parents and caregivers for poor supervision. But our Code of Patron Conduct will be useless if we are unwilling to enforce it. We must remember that the Code will not necessarily return us to the old days of monastic silence and discipline, but it could be an effective tool to decrease inappropriate behavior and enhance safety for all patrons.

As for Tanya, she'll survive the librarian's demand that she stop playing with that electric door. The Mother Goose program beckons, as does storytime and the summer crafts club. The latter will allow her to put her fingers to much better—and safer—use.

Shelving and safety: An overview

10

Shelving is fundamental to our culture. It is likely that prehistoric cave-dwellers stored their tools on any convenient stone shelf; millennia later, large shelving systems holding everything from clay tablets and papyri to pottery, grain, and weapons became common in the temples and warehouses of the early cities. Nowadays almost everything we eat, wear, and read has spent time on a shelf, and many of the materials used to build our residences, offices, and cars were stacked on shelves before they were hammered, welded, bolted, or screwed into place.

Since shelving is such a basic physical object, so common and apparently uncomplicated, librarians often take it for granted. We can work around miles of shelving for decades without paying much attention to it. It seems as solid and dull and immovable as the concrete in the foundations of our buildings. In an age of increasing budgetary cutbacks, however, when library facilities might not receive the same level of maintenance as before, shelving deserves more consideration. Like everything else, it ages. Without proper inspections and repairs, our shelving can become dangerously unstable. For the sake of safety, we need a stronger awareness of the shelving that we work around.

Since World War II, libraries have purchased shelving of all kinds from a growing number of manufacturers. A survey of the shelving systems in Canadian libraries would reveal many different brands of shelving installed in myriad ways. Think of the huge metal ranges in university and college stacks that become what a British engineer calls "catacombs of literate behavior." Many university stacks were installed in the 1950s and 1960s as postsecondary institutions expanded to meet the needs of the Baby Boomers. Some of these stacks were constructed from metal that manufacturers acquired as military scrap. It is possible and wonderfully ironic that the university shelf holding a strong collection of works on Buddhism once formed the bridge of a destroyer or the turret of a 6-in. gun. Libraries are the fruits of peace: we have beaten our swords into plowshares and our navy into free-standing bracket shelving.

Canadian public libraries contain the broadest variety of shelving. Newer public libraries have shelving systems, designed by teams of librarians and architects to provide convenient access to collections. As in any location where the weight of furnishings is heavy, floor loading is a constant concern. Library planners insist that the combined weight of all shelving and furniture should not exceed 150 pounds per square foot, and this figure can easily be respected as planners can determine exactly what will be installed or positioned in a particular space. Over time, however, librarians might modify a system to meet new needs. For example, in a large public library, several old ranges were removed from their original location to accommodate an expanding children's division. The old ranges came to rest in a storage area on a different floor. In light of the size of the public library and its inventory of furnishings, such a change might seem negligible. But the floor loading in its storage area has increased, and while floor collapse is unlikely, there is a chance that the floor will shift and cause a shelf collapse.

We refer to the configuration of our shelves as a system. This designation is accurate as long as the configuration is developed as a single entity for specific purposes and in response to the library's physical conditions and limitations. When we alter the system, move things to different locations, revise and reinstall, our system begins to lose its integrity. It becomes a collage. This is not necessarily a bad thing; many shelving collages serve patrons better than the original system. But reinstalling shelving introduces new stresses on bases, brackets, and braces. Wear-and-tear can accelerate. Consider how rickety many shelves become after we move them.

Since special librarians acquire many of their furnishings from various sources throughout their host institutions, special libraries are more likely to contain shelving collages. An extraordinary example is an engineering library in Vancouver, BC. The technical reference collection is housed in standard particle board units moored to the wall. The circulating book stock is kept in ranges of heavily braced commercial shelving that could support the weight of the entire collection, case shelving that once held a collection of fossils from the Burgess Shale, and old bracket shelving that might have been to sea in another incarnation. Beside the librarian's desk is the ready reference collection—phone books, dictionaries, AACR2, and a restaurant guide—housed in the shelving of a "home entertainment center" from the 1970s: this contains separate compartments for your television, bar, sound system, and Bee Gees albums.

Such a collage does not seem stable, but the engineers who use the collection inspect the shelving regularly to make sure that it poses no risk. The librarian reports any sign of wear-and-tear, any squeak, groan, or list. Repairs are quick. A shelving unit can be remoored, adjusted, rebracketed, or simply replaced in a couple of days. This nonsystem is not ideal, but it does its job inexpensively, efficiently, and safely.

A shelving risk analysis begins with an inventory of all shelving on the library site. This inventory should include the following:

- *The number of self-contained shelving units in different departments.* A self-contained unit can stand on its own without support from another structure.
- *The kinds of shelving on the site.* This section includes a description of the shelves in all public and workroom areas, as well as the old storage collage in the basement, and the wall-mounted shelves for cleaning supplies in the janitor's closet. Revolving cases, book trucks, compact shelving, file cabinets, and shelves mounted in workstations should be noted.
- *A brief description of how each unit is moored to the wall, floor, or other units.*
- *The estimated weight per square foot of shelving on different floors.*
- *The estimated age of all units.*
- *All floor plans, from the original configuration to the current layout.* These plans often contain valuable information on floor loading, vendors, and ages of units.
- *Any risk history.* Has there ever been a shelf collapse in the library? Has any staff member or patron ever been injured by a shelf? Is there any place in the vicinity of a shelving unit where injuries have occurred in the past?

The inventory leads into a shelving inspection record, which includes the following sections:

- *A description of wear-and-tear and any components that could cause injuries.* Note parts that are loose or missing, any listing or change of shape, rust spots, sharp edges, and any mooring that has weakened over time. One useful test for long ranges of shelving is the "one-hander": if you can cause a range to undulate by pressing against an end-panel, then it is probably not safely moored.
- *Janitor's shelving.* Is it securely mounted? Does it hold potentially dangerous chemicals, such as ammonia? If so, are the containers securely stored? Is there a lip along the edge of the shelf to prevent containers from falling off?
- *Revolving cases.* Have they been spun too many times by small children waiting for browsing parents? Does any case seem inclined to tip? When you turn a revolving case, listen to it. Does it sound as if one of its parts is rubbing against another? Ideally, you will be able to inspect your revolving cases when they are fully loaded, since greater weight reveals weaknesses faster than a light load.
- *Book trucks.* Are they inclined to tip as you turn a corner? Are the wheels loose or worn out? Does any truck feel unstable when fully loaded?
- *Compact shelving.* Does it run smoothly along its tracks? If your compact shelving system requires electricity for movement and monitoring, is the wiring in good condition? Does the shelving squeak or rattle as it moves? Has the vendor or physical plant manager inspected your compact shelving lately? Compact shelving is rumored to be a greater risk to users, but media reports do not confirm that well-maintained compact shelving is any riskier than other kinds of library shelving. In fact, state-of-the-art compact shelving can be much safer than other systems that are unmoored or cheaply constructed. Nevertheless, since compact shelving comprises moving parts and, in the more sophisticated systems, electronic circuitry, it must be inspected and maintained more often than other systems.
- *File cabinets.* Are they securely moored to the floor or wall? Are the drawers easy to open?
- *Workstation shelves.* Are these overloaded with software manuals and unmoored printers? Are they strongly moored to their supporting components?

Other risk factors that you should note during your inspection are as follows:

- *Regional risks.* The most important regional risk for shelving is an earthquake. British Columbia has what many seismologists believe to be the highest earthquake risk in Canada. Southwestern British Columbia—where the largest libraries in the province are located—could experience a quake many times stronger than any that have shaken California in recent years. Unmoored shelving of all kinds could collapse in even a moderate quake.

A common risk that is not life-threatening is moisture, which seeps into walls and can deteriorate wall moorings. Libraries located in regions that regularly experience heavy rainfall should note any sign of leaks or seepage, such as stains on the floor or walls. Fortunately, moisture is much slower to loosen furniture than an earthquake; nevertheless, its cumulative effect over several years should not be ignored.

- *Type of library.* Shelving risk increases with the number of patrons and staff members who spend time in and around shelving. Busy libraries will inevitably experience more wear-and-tear to all furnishings and fixtures than special libraries with limited access.
- *The library building.* Older buildings are more likely to have suffered more structural damage over the years, and records of this damage might have disappeared. Thus, an older

building, such as a Carnegie library, might have less stable flooring and foundations. Walls in these libraries might not be able to provide effective mooring. While each building must be considered on its own merits, newer buildings constructed according to the specifications of the current National Building Code are less likely to experience shelving safety problems, especially if shelving systems are an integral part of the building's design.

- *Aisle widths.* Generally, the wider the aisles, the less risk there is from shelves. Library administrators are determined to use their shelving space as cost-effectively as possible, but shelves situated too closely together can result in more wear-and-tear to shelves and collections as book trucks bump their way along narrow aisles. Narrow aisles also encourage staff members and patrons to bend and twist themselves in such a way that back injuries can ensue.

A comprehensive risk analysis should include suggestions for risk mitigation. Often, mitigation measures to reduce shelving risk are too expensive to take in the immediate future. In such instances, a phased mitigation plan is recommended. It is wise to deal with the most unstable shelving first. If it cannot be repaired or made more secure, it should be replaced. Those unstable units closest to public and staff areas should be repaired or replaced first. Note that in many libraries, the riskiest shelving is in the reference area. The reference shelves hold greater weights than other shelving units, and they are in constant use by staff and patrons. Frequently, a set of the heaviest reference shelves stands directly behind the reference desk. If these shelves are deemed unsafe, they should be repaired or replaced as soon as possible.

Are shelf collapses common? It is impossible to judge from media reports. What is important in such reports is not what is discussed, but what is left out. In fact, it is rare to hear the reasons for a shelf collapse, especially if somebody has been injured or killed in the event. But certain circumstances usually prevail: the shelving that has collapsed is older, heavily loaded, and not well moored, the victim has worked around the shelving for years, the victim was working alone at the time of the collapse, and did not receive prompt help.

Most shelf collapses, however, are not fatal. In this writer's experience, they occur in basement and back-room storage areas, frequently during off-hours. The shelving is either old wood or even older metal, and the flooring in the immediate vicinity appears uneven. A risk analysis would have easily confirmed the instability of the shelving before it collapsed, but storage shelving is not considered a high priority. Unfortunately, as budgets shrink, safety for entire libraries is becoming a less outstanding issue to administrators, who must rationalize every small expense. We trust that they will reconsider their priorities before gravity strikes a blow in a crowded public area.

"It's not just the books!" Wheelchair patrons speak out

11

Chess masters of the last century relied on defensive strategies to lure their opponents into error and defeat. At present the brutally aggressive player is rare. Carl, aged 40, is an example. On a late summer day in the activities room of a BC public library, we find him destroying simultaneously three members of the local chess club. Carl arranges their captured pieces in tidy rows to one side. The local club never had a hope. Once upon a time, neither did Carl.

Every year, accidents kill and maim numerous loggers in BC's forests. Carl lost his legs when a 20-m Douglas fir fell across them. He awoke in a Vancouver hospital with two neatly bandaged stumps. An orderly taught him the basic chess moves; a girlfriend gave him a pile of texts on the middle game and the tactics of famous Russian players. Carl studied and practiced endlessly. Within a few weeks, he had whipped a platoon of orderlies and nurses, his physiotherapist and rehabilitation coordinator, and the hapless orthopedic surgeon who fitted him for a wheelchair.

Carl has become a regular library patron for several reasons. First, libraries contain much of the information he needs to complete his vocational retraining program. Second, they offer handicapped stalls in the washrooms. Third, they provide quiet space for chess games. Lastly, they're full of willing victims, people who will sit with Carl for hours and allow him to tear them apart, piece by piece.

11.1 Safe spots

"When you end up in a wheelchair, you look for safe spots, places where you can be comfortable," says Carl. "You're vulnerable, even in cities that try to accommodate the physically challenged. But libraries are almost always safe. People won't push you out of the way or ram you with a shopping cart, and library staffers are helpful. Most wheelchair people will tell you that libraries give them a break from the anxiety of dealing with sidewalks and malls."

Many large North American libraries started to adapt their facilities for wheelchair access in the early 1960s. Forty years later, most public libraries have parking spaces reserved for handicapped persons. A wheelchair patron can enter the library either through a street-level door or along a ramp that's not too steep. Ramps have handrails on one or both sides. Doorways and aisles are wide enough for wheelchair patrons to negotiate without difficulty. Elevators remain open for several seconds, enough time for a wheelchair patron to enter or exit without colliding with doors in motion. Those elevators are serviced regularly to ensure that they stop at a point level with the floor, so that wheelchairs don't have to lurch dangerously when they cross the threshold. Signage is geared to meet wheelchair patrons' needs, especially in directing them to washrooms and emergency exits.

Robertson on Library Security and Disaster Planning

11.2 Library attitudes

Attitudes toward wheelchair patrons have become more sophisticated as well. Library staffers can anticipate problems that a wheelchair might encounter in different parts of the library, and assist patrons at the appropriate moment.

"Librarians have learned patience," says Mary, an 88-year-old retired landscape architect with a balance impairment who depends on a wheelchair to perform errands. "They understand that some of us can't get past turnstiles and stairways, and that we can get stuck turning a corner or changing direction on certain types of carpets. Some of us simply run out of energy and need a hand."

Mary is a heavy reader, and can't conceive of life without access to books and magazines. She relies on Vancouver libraries for her chief source of entertainment and her main connection to the world at large. Not even her TV can give her what she gets from a stack of old magazines or a weekend's worth of mysteries. Since she started to use a wheelchair, she has become acutely aware of the different ways that library staffers treat her.

"There's a subtlety that many librarians develop," she says. "They take note of people in wheelchairs. When a wheelchair patron has trouble reaching a book on a high shelf, they appear to offer assistance. But they have a gift of not making you feel self-conscious about needing help. That's a special kind of service. It's not available in every library, but it should be."

Many wheelchair patrons do not like excessive attention. They do not like staff members to follow them around and monitor their every move.

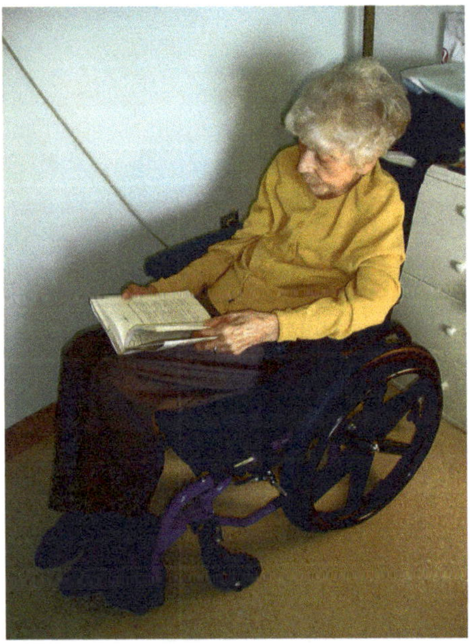

11.3 Independence on wheels

"It's annoying when all eyes are trained on me," says Carl. "Once I'm past the front door, I don't want extra attention. I get irritated with the non-handicapped people who monopolize Internet stations that are set up for wheelchair folks, but I don't want the librarian to intervene on my behalf and reprimand some character who wasn't aware of me because he was too busy examining a popular website. I can handle those situations myself. I simply say, 'After you,' and the guy will usually make way for me. But when the librarian steps in, he gets embarrassed, and I feel like a nuisance. This is not a small point. When the librarian or any other service worker acts on my behalf like that, I feel like a cripple. I assure you that I can handle most of these situations myself."

Librarians must realize that every wheelchair patron is unique, and that some occasionally need a special kind of help. Liz, aged 38, is one such. She suffers from a form of multiple sclerosis that allows her to walk with a cane but occasionally forces her into a wheelchair. She and her family recently moved to Vancouver from Northern Manitoba so that she could receive more advanced medical treatment and counseling.

"My MS is unpredictable," she says. "I have good and bad days, but I can't alter my routine significantly, not with two daughters in elementary school. Our schedule includes regular visits to the local library for storytime, the young readers' club, and random browsing. I think it's important to make the library part of my daughters' lives, so no matter how unstable I'm feeling, we're off to the library. With my husband at work, I'm in charge of the girls."

11.4 When to ignore the rules

Library policies and practices may limit the amount of physical assistance that library staff members can offer Liz, but she admits that occasionally they have ignored official restrictions to help her.

"Sometimes I'm feeling strong enough to get about with nothing more than a cane," she says. "But by the time I reach the library, I'm tired. That's when I'm at my worst, and there's a risk that I'll collapse. At my library branch, however, they look out for me. One of the staffers appears out of nowhere and helps me to a chair. Then somebody brings me a glass of water, while the children's librarian grabs my girls and gives them something to read. When I'm feeling better, the librarian makes sure that I have a lift home. I guarantee that there's nothing in the procedures manual that demands this level of attention devoted to one patron. My husband says that I probably receive better care in the library than I do at the hospital."

When Liz appears at the library in her wheelchair, the staff treat her with the same consideration. She notes that an assistant has volunteered to "escort" her around the library.

"That means that he will push my wheelchair around the stacks, and hand me anything I can't reach. He'll also check out my books and pack them in the bag at the back of my chair. But he doesn't treat me like an oddity. We enjoy complaining about the weather and civic politics, and we wax indignant about the incompetence of local hockey and football teams. In other words, there's nothing strained or uncomfortable about my interaction with the library employee. And that's how it should be."

11.5 Individual respect

Carl, Mary, and Liz are encouraged by the ways in which libraries serve them as individuals.

"You have to be impressed when you see librarians getting to know you and your physical limitations so quickly," says Carl. "After my accident, the social worker sent me to a support group for people who were recently handicapped and who relied on wheelchairs. What struck me right away was that wheelchair people have different limitations and challenges. There are paraplegics and quadriplegics. Some of us are incontinent, or asthmatic, or depressed, or heavily medicated. Elderly wheelchair users can get drowsy and fall asleep in the reference area. Younger users might require constant supervision. But you don't hear librarians complain about us. We seem to be as welcome as anyone else, no matter how severely challenged we are. And that's why so many of us become regular library users. It's not just the books, it's the quality of the staff and the way they treat us like ordinary people."

Liz mentions that she sees an increasing number of wheelchair patrons in her library. She notes that all have different ways of using libraries.

"For example, my husband Tom has just joined a chess club, and the fellow he's playing tonight is in a wheelchair. He's supposed to be a good player. Tom's looking forward to meeting him."

Poor Tom. Carl has no mercy.

Beyond band-aid solutions: Dealing with medical emergencies in your library

12

Twenty years ago, you took a one-day training program that taught you how to deal with different kinds of fractures, burns, and lacerations. Perhaps you learned cardio-pulmonary resuscitation and the Heimlich Maneuver. You hung your first aid certificate on the wall with pride.

Years passed. Nobody around you suffered any dangerous gashes or broke any bones. Or if they did, paramedics appeared so quickly that there was no opportunity to demonstrate your skills with splints and bandages. Bloody noses dried up before you could stanch them. Kitchen scalds dived under cold taps before you could soothe them. Admit it: as a first aid provider, you've been redundant from the start.

But things can change quickly. Tomorrow, during your shift at the reference desk, one of your library's regular patrons will arrive with a headache that aspirin isn't helping. Her name is Anne. She's 70 years old, a smoker, and overweight. She scans her favorite fiction shelves, and suddenly can't read the titles on the spines. One side of her body becomes weak; she has difficulty in standing. She has suffered a stroke. Could you recognize the signs? Perhaps not. Your first aid trainer didn't discuss strokes (which are sometimes called "brain attacks"). There wasn't time during that training program. But now you have Anne on your hands. Are you ready to provide the first aid that she needs?

12.1 Competent response

In fact, even the most thorough training will not necessarily make you a competent first aid responder. As in other areas of endeavor, some people are better at dealing with real emergencies than others. You may have read about the warning signs of strokes, but will you recognize them in Anne when she tells you that she doesn't feel well?

Fortunately, serious medical emergencies in libraries are infrequent. Anecdotal evidence from across North America indicates that public branches serving large numbers of elderly patrons are more likely than other libraries to experience emergencies requiring quick medical response. Most staff injuries are due to falls, inappropriate lifting techniques, and the sharp edges of scissors and broken metal shelves. Occasionally an employee takes time off to nurse a bad back or a sprained wrist. Disinfectant and bandages will be all that's necessary for minor cuts. Libraries are not coalmines or heavy manufacturing plants, and the injuries that we see in the Cataloging Department are not nearly as ugly as those we see in an open pit mine or a steel mill.

Robertson on Library Security and Disaster Planning

Hence we're content to muddle through, and we accept our lack of preparedness. Even when we hear about a patron at another branch who complained of chest pains and ended up in hospital for a coronary bypass, we don't become alarmed. In that case, the patron called 911 on his cell phone, and shortly thereafter an ambulance arrived at the main entrance and took him away on a stretcher. The library staff didn't know that there was a problem until they saw the ambulance leaving. Those cell phones are wonderful, aren't they?

You won't be so lucky with Anne. She's in bad trouble, as blood from a ruptured vessel in her brain seeps into surrounding tissues and causes the pressure in her skull to increase to dangerous levels. If she doesn't receive treatment quickly, she could be paralyzed. Or she might die. She has no cell phone, and is now disorientated and frightened. She asks another patron for help; he assumes that she's drunk and ignores her. She holds herself up on the edge of your reference desk and tries to tell you that she's feeling sick. Her speech is garbled. What should you do?

12.2 First aid policy

At any time during business hours, library employees might encounter people in Anne's condition, or with another kind of dire medical problem that demands a fast response. How you deal with such circumstances depends on a number of factors. First, does your library have a first aid policy? If so, do library managers and employees take it seriously? You're familiar with numerous policies that managers draft, file, and forget. And just because a policy is available online does not mean that anyone will read it. For example, you're supposed to change your computer password every quarter. That's a standard operating procedure in your library's hardcopy manual and on the website. Have you changed your password lately?

Your first aid policy should require every library to have properly trained first aid attendants on site during business hours. Each branch should have the appropriate number of first aid kits stored in easily accessible places. The larger the staff at a particular site, the greater the number of attendants and kits that will be necessary. High-traffic sites should have more first aid supplies. How much? Your local Workers' Compensation Board or fire department can recommend the amount of supplies that a particular branch needs. First aid equipment vendors can also suggest the size of kit that would be best for your staff and patrons.

First aid policies should also ensure that it's wise to dial 911 if there's any doubt about an employee or patron's safety. This is common sense, but as such, it's often forgotten or ignored. Thus while a middle-aged man lapses into a diabetic coma in the washroom, librarians conduct an online search for their library's first aid policy, to determine the right course of action. Unfortunately this isn't fantasy. These events happen more often than we think. They lead to questions from ambulance crews such as "Why didn't somebody call us sooner?" and "Didn't anybody know about 911?"

12.3 Attendants and kits

Second, how effective are your first aid attendants? Is their training up to date? Do they meet at least twice a year to review circumstances that might lead to changing first aid needs, such as growing numbers of elderly and disabled patrons? A certificate does not necessarily make a proficient first aid attendant. He or she needs occasional refresher training, and an awareness of the potential problems that could occur at a specific site.

Third, your library bought first aid kits for each branch. Do you know where yours is? Having to hunt for a kit during an emergency is unacceptable. A good practice is to mount the kit on a wall in a central place such as the staff room or kitchen area. Once it's installed, employees should respect it. A kit should not be used as a source of gauze bandages for the librarian's Halloween mummy costume. The cotton swabs and rubbing alcohol should not serve as cleaning equipment for keyboards. That rolled-up bandage is intended to bind sprained wrists and ankles, not to hold together the old felt board or puppet theater. Your kit deserves its proper place and appropriate use. Otherwise it will not be useful in an emergency.

Finally, in a case such as Anne's, you must recognize when a kit and even the best first aid training will not be enough to keep a sick or injured person safe. To help somebody who's in the most life-threatening condition, you must be able to recognize the problem and call for professional assistance as quickly as possible. These conditions include various kinds of strokes, heart attacks, drug overdoses, allergic reactions, and injuries involving substantial blood loss or serious burns. Your ability to recognize the nature and severity of the problem could be essential for the sufferer's survival. You may assume that it's easy to spot the signs of a stroke, but unless you've reviewed these signs carefully, you might not be ready to identify an actual case.

12.4 911: Don't hesitate

When you dial 911, the operator will ask if you need police, firefighters, or an ambulance. Ask for the latter. The operator might ask you to describe Anne and her condition. No one expects you to give an accurate diagnosis of Anne's problem. You're permitted to say that you *suspect* that she has suffered a stroke, and to mention why you think so. If your assessment of a person's medical condition is incorrect, you shouldn't expect a rebuke from the emergency ward doctor. It's easy to misinterpret the signs of different diseases and conditions. Your first aid policy should contain a clause that advises you to dial 911 without hesitation. If the subject of your call turns out merely to be drunk, ambulance attendants will still make sure that he or she is safe, and might provide a lift to the nearest emergency shelter.

After a medical emergency at your branch, your senior administrators will ask for a report. Use this opportunity not only to describe what happened, but also to recommend improvements to your on-site first aid capability. You may note that you need a refresher session for your branch attendants, and ask for professional trainers from the

Red Cross or St. John's Ambulance to deliver it. Request a new first aid kit if you need it and advise your administration that other branches in your system might benefit from the same upgrades.

With luck and a good emergency medical team, Anne could return to your branch soon. She might walk with a cane and have trouble pronouncing her words clearly, but you'll be pleased to see that she's recovering. Treatments for strokes have improved over the past decade, allowing many sufferers such as Anne to resume most of their normal routines. She'll be scanning those fiction shelves again, which is always a good sign.

13 Not for the squeamish: The dirty truth about your library

Keeping your library clean is a continual challenge. You cannot ignore the grimy facts, unpleasant as they are. Consider the sources of dirt in your library before you try to improve your library's sanitation. Fortunately, there are effective methods of dealing with the issue.

Filth abounds in your library. Consider the fiction shelves, which are crammed with D.H. Lawrence, James Joyce, Henry Miller, and Anaïs Nin. Your self-help section contains an abundance of sex manuals. The Children's Department is especially grubby. Just look at those books by Shel Silverstein and Francesca Lia Block.

In fact, titles by notable authors such as these are handled more often than most library items, and frequent handling leads to heavier soiling. Readers sneeze in the gutters of great literature and spread their bacteria across the worldwide canon. They would shudder to think that they spray their favorite works with droplets of mucus, but they do. And they add to a book's deterioration when they store it on dusty bedside tables, on unswept floors and unvacuumed carpets, and even in the grotty cavern under their beds.

13.1 Caldecott readers and pets

Children are to blame for some of your library's sanitation problem, since they have yet to master basic hygiene. Oblivious of the concerns of parents and other adults in proximity, they cough and sneeze and explore their noses with tiny vigorous fingers that subsequently find their way into the latest Caldecott winner. Before they leave the library, they visit the washrooms and demonstrate a lack of aim that frustrates even the most hardened janitor.

Arriving home, they coat the pages of a library book with Grandma's strawberry jam, which stiffens so as to ensure that nobody can open the book without tearing the paper. Children can also add multicolored stains that would defy identification in a forensic lab. What's brown and red, and smells like an ancient cheese? (Answer: ancient cheese stored in an aforementioned grotty cavern.)

Household pets are notoriously unclean. Rover, a big loving brute, slobbers all over everything, and leaves his hair and dander between pages, in video cassettes and DVD containers, and on the delicate surfaces of CDs. Moggy the Siamese cat is tidier, but she can distribute almost as much hair as Rover.

13.2 Guilty as charged

The chief offender, however, is usually an adult.

"Kids and pets take the rap for a lot of damage and dirt in library items," says Carol, a Vancouver children's program coordinator. "I'm sure that they can be messy, but adults are responsible for the worst damage to books, and that's true in most libraries."

She refers to a business motivational video that a well-dressed gentleman returned to a branch earlier this year. The video stank of beer. The gentleman explained that his infant daughter had accidentally overturned his pint on the video.

"I didn't believe it," says Carol. "I think he blamed his child to get sympathy, so that we wouldn't charge him for a new video. I hate to admit it, but librarians fall for this sort of fib all too often. We are within our rights to tell people that they are responsible for their children, and that parents must pay for the havoc that toddlers wreak. We can't afford to let people get away with it."

13.3 Back to you

Eventually patrons return soiled items to the library. Every day, dirt in different forms flows in through the front door and return slots. Ventilation systems allow pollution to mix with a library's indoor air, and particles from motor vehicle engines settle on every indoor surface along with other dust. Moisture ingress causes mold to form along the edges of windows, and in wall and ceiling cavities that are almost impossible to clean.

But human beings are the source of most sanitation problems. No matter how careful you are in bathing and grooming yourself, you carry germs on your skin. You breathe out contagions and deposit viruses on every surface you touch. Also, since you eat, you excrete. This bodily function is not the sort of topic that appears on library school syllabi, but it is a matter of serious import in facility planning and design. The wrong choice of washroom layout or fixtures can lead to poor sanitation and higher cleaning bills. For example, trash receptacles should be large enough to contain used paper towels and anything else that patrons discard in washrooms. This is common sense. But often libraries install diaper-changing tables in washrooms without increasing the size and number of trash receptacles. The result is horrible, as receptacles overflow and the odor of soiled diapers wafts through public areas.

13.4 Food service

"There's probably a correlation between the availability of food in a library's immediate vicinity and the frequency of its washroom use," says Brenda, a Toronto technical services librarian. "The situation can become especially unpleasant if there are multiple coffee shops in the area, or worse, pubs. Not only do you see a lot more litter on library property, your washrooms will be messier."

Brenda suggests that the trend toward selling food in libraries is "a nice idea," but that it will probably lead to lower sanitation levels and increasing cases of insect and rodent infestation.

"Rats are always a potential problem, and now we're expected to serve them coffee and muffins," says Brenda. "Assuming that patrons won't drop bits of food on floors is naïve. It's all very well for libraries to offer food service to patrons, but I doubt that many managers have considered the ongoing cost of pest control. Add moisture, and you'll find silverfish under the furniture. And then there are cockroaches. Have you ever dealt with a patron who has witnessed a roach scurry across a table in a reference area? Sometimes they're furious. It's as if the library has insulted them with poor sanitation. Infestations can contribute to poor employee morale, too. Are those muffins in the foyer really worth all the grief?"

13.5 No respect for owls

Urban librarians are also concerned about the effects of pigeon populations on sanitation. Pigeons appear harmless, and people enjoy feeding them stale breadcrumbs on pavements outside libraries. But pigeons defecate freely. Flocks can deposit large amounts of acidic dung on roofs and windowsills, and around doorways. On a hot day, the resulting smell can become overpowering. Removing the stains is difficult and costly, and the question remains of how to discourage pigeons from returning.

"You can't shoot them," says Jerry, a Vancouver special librarian who has dealt with pigeons on his windowsills for years. "You can try to frighten them with loud noises, or you can install a fake plastic owl in their congregation area. But these measures aren't effective for long. Our owl was covered with pigeon droppings a week or so after we installed it. Any solution will be temporary, including wire netting and other barriers. Even if you were permitted to cull pigeon populations with a shotgun, the survivors and their descendants would eventually come back. Attempts to get rid of them often don't work. In my library, the best thing we can do is to clear away build-ups of droppings to control the smell."

One small consolation in a world of unmovable wildlife: bookworms are rarely active in Canadian libraries. The cold climate does not allow them to thrive. Rare book departments hold numerous manuscripts and printed volumes scarred with wormholes, but living worms are uncommon. Should you discover evidence of a living worm, you can destroy it by wrapping the volume that it inhabits in heavy-duty plastic, and leaving it in a freezer for a few hours.

13.6 Pandemic influenza

The complexity of the sanitation problem can overwhelm you. There seem to be too many angles to consider, and too many tasks to complete just to keep your library at an acceptable level of cleanliness. Moreover, there are new aspects to take into account. For example, pandemic influenza. It might turn into a mass killer, or remain nothing more than a nuisance. But you are expected to prepare your staff members and facilities as much as possible for any kind of outbreak, deadly or mild. You must at least reduce the likelihood of contracting influenza in your buildings.

You should develop procedures to maintain good sanitation. Acknowledge the specific sorts of bad sanitation that exist at different sites, and realize that in many cases you can do no more than reduce the level of dirt, and not eradicate it.

13.7 Helpful measures

First, make sure that your janitorial services—in-house or contracted—are adequate. Maintenance staff should have their responsibilities clearly documented. It is not enough to ask them to clean a washroom. A set of specific tasks in the form of a checklist is the best way to describe cleaning duties.

Frequent—at many sites, daily—cleaning is advisable for objects that people touch in high-traffic areas: handrails, elevator buttons, telephone handsets, photocopier keys, and keyboards. This measure is especially important during outbreaks of influenza.

Second, ask patrons regularly and firmly to take good care of the items they borrow. On posters, brochures, and websites, warn them about the risks of handling foodstuffs near library books and other media. You need not be stern. In fact, a light-hearted request to patrons not to sprinkle the latest biography of Conrad Black with caviar and champagne is more memorable and effective than a plain demand for proper care of library property.

You should gently remind children that library books are wonderful things, and that all of us must keep them away from jam, peanut butter, bodily fluids, and Mr. Underbed. Everyone should strive to restrict Rover and Moggy's access to print, unless it be newspaper used for protective lining.

13.8 Hand washing

Third, post signage regarding hand washing in all washrooms, including those for employees. Provision of adequate supplies of soap and towels is necessary, along with hand cream for employees. Antibacterial soap is not essential. Ordinary soap used with warm water will reduce bacteria on hands just as well as most antibacterial products. What is important is employees' and patrons' willingness to wash their hands thoroughly as a matter of course.

Feel free to advise everyone in your library not to cough and sneeze without covering mouths and noses. If people use paper tissue, they should deposit it in a trash receptacle, and not in an empty space on a shelf. Do not tolerate spitting anywhere on library property.

Fourth, if your library sells food, ask patrons to deposit napkins, sandwich containers, paper cups, cans and bottles, and other trash in a clearly marked receptacle, which should be emptied every business day. Libraries with larger food counters or cafeterias should arrange regular (i.e., quarterly) inspections with a pest control vendor.

13.9 Light and space

Fifth, regarding all facilities in the planning stage, either for new construction or renovation, architects and library managers should consider long-term sanitation issues, particularly in busy public areas. Washrooms must be well illuminated, with enough space to maneuver in cubicles and in urinal areas. The number of sinks and trash receptacles should be appropriate to the size of the building and its human traffic. Acceptable washroom design should not be taken for granted.

Remember that sanitation is a constant concern and not something that you can attend to once a year and forget the rest of the time. People will not stop coughing and wheezing and treating the washrooms carelessly; nor will they always use trash receptacles as they should. What is clean one moment is filthy the next, as a small child grabs a handrail, or her parent pounds a keyboard in the reference room. Anything that they—and you—come into physical contact, will get dirty. You may say that *Lady Chatterley's Lover* is nothing to sneeze at, but inevitably readers will demonstrate otherwise.

Senior security: Keeping the elderly safe

14

Hilda is confused. She had a question for the reference librarian in the branch near her seniors' residence, but it has slipped her mind. She tries to remember. Was it to do with the catalog? Hilda's memory is poor owing to dementia. She is 90, and jokes about not being able to recall her own name. Secretly she worries that one day she will indeed forget it, along with everything else. At present, however, she'll struggle to come up with that wretched question, which concerned the catalog. Maybe.

Meanwhile, Charles is not content to work things out by himself. He demands help immediately, and bellows at the clerk at the circulation counter. Charles has a form of dementia—possibly Alzheimer's—that is different from Hilda's. He has angry and sometimes violent outbursts, and has assaulted a care aide at his residence. He makes library employees nervous. "The paper's in front of you, Charles," says the clerk, a young woman who knows how to calm him down. "By the way, your glasses are on the table in the quiet area, where you left them." Charles shuffles away sheepishly, newspaper in hand.

14.1 Physical threats

In fact, the risk of workplace violence caused by seniors is rare in libraries, despite Charles's bad behavior. Moreover, there are few reports in North America of elderly villains stealing computers or spray-painting graffiti on library exteriors. Seniors are usually well behaved. What concerns security specialists are potential physical dangers to seniors on library property, from muggers and anyone else who would attempt to intimidate, assault, or rob an elderly person walking to a bus stop or car.

"You get some nasty characters hanging around," says Dan Henderson, a security consultant in Vancouver. "I'd say that the parking lots are the worst places for seniors, who are easy to knock over and rob. It has happened before, and as the population ages, I think we'll see more of these crimes."

Henderson suggests that the riskiest locations for seniors are malls and any high-crime areas through which seniors must travel with purses, groceries, and other bulky items. Thieves rely on old tricks such as volunteering to carry heavy bags to a senior's car, then running away with the bags and their contents. In some cases, the victims are so embarrassed that they do not report the crimes to the police.

"Old folks are also disinclined to complain to librarians about getting ripped off," says Sara Thomas, a librarian in Florida who has worked in Toronto and Winnipeg. "I know of cases where thieves have pulled the carry-your-bag scam, and stolen library books in the process. In a couple of cases, the oldies offered to pay for the books rather than reporting them as stolen. They said that they didn't want to make a fuss."

14.2 Watch yourself, son

Not all seniors are soft targets, however. Some years ago, a thief on a skateboard attempted to wrestle a bag of library books away from Jim Edwards, a retired credit union manager in Toronto. Jim pushed the thief, a teenaged male, off his skateboard, and then grabbed back the bag.

"I told that kid to cease and desist," says Jim. "I told him that I was at D-Day and I knew how to handle myself." Unfortunately, the thief will not necessarily stop threatening seniors. It's possible that in future he will simply choose weaker victims. Obviously, not many seniors are as prepared as Jim to defend themselves. Should libraries develop security systems to protect seniors in particular? Standard measures including guards, CCTV, and patrols of public areas might be effective in warding off villains in general, and not only those who would steal Jim's book bag. A "seniors focused" security system might not be needed. If, however, a library attracts an above-average number of elderly patrons, there are simple ways to enhance their safety. For example, some libraries post large signs warning patrons that thieves are active in the library. These signs not only warn older patrons about the risk, but also deter thieves and anyone else who might be tempted to commit a crime in the library.

14.3 Escorts and lighting

In neighborhoods with high crime levels, seniors who must walk to a library parking lot or nearby bus stop could be escorted by a library employee. This measure does not consume much time or energy, and can discourage muggings. It is also wise to increase the lighting in frequently traveled exterior walkways, parking lots, and exits, since villains tend to avoid well-lit areas.

Above all, library employees should be trained to spot possible trouble before it occurs—for example, a person who loiters in the library and appears to be more interested in elderly patrons than anything else, or hangs around outside entrances and exits, and watches patrons as they enter or leave the building. Persons who loiter in or around washrooms also deserve the librarian's attention.

"Instinct might tell you that a certain character is up to no good," says Sara Thomas. "I'd advise library employees to trust their instinct. It's not always right, but it can anticipate criminal acts. Most experienced librarians, and especially those who have worked in city cores, know what I'm talking about. The bad guys often look as if they're up to something. And you can hinder them by letting them know you're watching."

14.4 Nice toaster

Librarians are not as forthcoming on the topic of the damage that seniors might cause to library property. This is because it is customary to let them off lightly. Seniors are not as strong and as well coordinated as they once were, and they drop books on the floor and in the bath and everywhere else. On countless pages, seniors have poured tea or coffee or perhaps a little whisky; they have moistened magazine covers with pet food, and dried CDs in toasters after pouring juice on them by accident. ("I didn't think that a hair dryer would do the job as good as my toaster oven, which usually treats everything so nicely," said an apologetic great-grandmother to a librarian in Western Canada.)

Equally common is seniors' loss of books. Visit the lost-and-found department of any mall or department store that attracts large numbers of old people, and you will often discover dozens of library books that were unintentionally abandoned in the cafeteria by elders looking for half a sandwich and a pot of tea. Seniors leave books behind in hospitals and community centers, and on park benches. They don't remember where they left those novels and biographies; sometimes they assume that they have returned them to the library, and become indignant when librarians tell them that there's no record of the books' return.

14.5 Computers are safe

"You have to accept a higher rate of loss from collections when you're dealing with older patrons," says a librarian in Oregon, who asked to remain anonymous. "There's not much that you can do with people who have failing memories. It's not as if they're malicious or dishonest. I don't like to charge them if they're regulars in the library, and occasionally they find the items they've lost and return them. I wouldn't feel comfortable in suspending borrowers' privileges in cases where a long-term patron has lost a couple of books."

One aspect of seniors' library use that surprises many administrators is their gentle treatment of computer equipment. Either they avoid it altogether or use it with great care, asking for assistance if something doesn't work for them. And some of them—dubbed Silver Surfers—become adept at database searching and Internet use. They may learn computer functions more slowly, but many seniors can pick up the required skills as thoroughly as younger patrons.

Frequently more worrisome than other security issues involving seniors are the personal and individualized challenges that somebody such as Hilda experiences. Although she has been a regular patron of her branch for decades, now she can become confused about the location of the washroom, exits, and her favorite spot in the reference area. She resists even the smallest change in her environment. She prefers to sit in a particular chair and gets upset if somebody is occupying it when she arrives. Sometimes she loses her bearings, but doesn't create a stir. Instead she wanders through the stacks for lengthy periods, looking lost.

14.6 The errant card

Hilda has misplaced her library card several times and has asked for replacements. When a care aide in her residence finds her misplaced card and returns it to her, she attempts to borrow items with it. But it has been canceled, and she must hunt through her enormous, ancient purse for the newer card. While this may seem like a minor difficulty to librarians, it is a major concern for Hilda, who is ashamed of what she calls her "incompetence." Accustomed to her, circulation desk employees will take the time to reassure her that Armageddon does not impend owing to a misplaced card. Hilda will also check out books, return to her chair, read them for a while, and then leave them in the library when she departs. Clerks reshelve the undischarged books, and the result is confusion in the circulation department. There are no guaranteed methods of dealing with such problems effectively in all cases. What works best is patience and familiarity with seniors' difficulties in managing matters that younger persons handle without thinking.

14.7 A little help from a colleague

Today, however, Hilda has succeeded in recalling her question for the reference librarian. It concerned the catalog, which Hilda enjoys examining at length. It is very different from the card catalog that she worked with when she was a librarian at universities on the Canadian Prairies. That was many years ago, before AACR and myriad technical advances. Now Hilda has heard of something called RDA, and wants to know what it involves. She asks the reference librarian, who groans. He admits that he doesn't know anything about RDA, but he'll find Hilda some articles on the topic.

"And maybe you'll be able to explain RDA to me, because I find it confusing," says he.

Hilda returns to her chair feeling useful and pleased with life. Some days are bad, especially when she's not sure where she is or where she has left things. Other days are splendid. This is one of them, thanks to RDA. She's determined to learn about it, and prove to the reference librarian that she's a good teacher. What are retired colleagues for?

Stealing from library patrons: A helpful introduction

15

Bank robbery is not what it used to be. These days, hold-ups produce paltry amounts of revenue, as tellers no longer rely on cash drawers crammed with wads of bills. Now behind tellers' wickets one sees mechanical cash dispensers that fear no robber with a cheap handgun. Often robbers flee from bank branches with a few hundred dollars or less. Such negligible swag is not worth the effort, especially when the police are diligent in their pursuit of criminals, and judges deliver harsh sentences.

There are much safer and more profitable ways to steal. Embezzling money from one's employer is increasingly simple in an age of electronic accounting, and illicitly diverting funds need not leave an evidence trail. Cheque fraud and forgery remain common. And of course there is always the straightforward theft of money and personal property as opportunities arise. The question is, what is the straightforward thief's best *modus operandi*?

Burglary is one option, although it can be fraught with risks: alarms, dogs, surprised and potentially violent householders, observant neighbors, broken glass, slippery stairs, and police officers who arrive quickly at the scene. Pocket picking in busy urban settings has a Dickensian charm, but in an age of tight trousers its fast detection is almost inevitable. Purse snatching is déclassé and favored by only the most desperate criminals, who rarely earn much, and are often caught. You are wise to prefer locations and techniques that are calm, quiet, and frequented by unsuspecting persons engaged in activities that absorb and distract them: for example, conducting research on financial markets, checking databases, or reviewing the works of Tolstoy, Shakespeare, and Dr. Seuss.

15.1 Avoid the rabble

Your local library is an ideal spot for theft. For centuries, librarians have worried about book thieves, and struggled to frustrate them with everything from threats and curses to vigilance and piercing stares. During the past 40 years, Tattle-Tape and RFID have become popular protective technologies—not only for books, but also for any other media that might be stolen. But you are not a book thief; nor are you eager to steal DVDs. You are hunting bigger game. After all, with the cash in one stolen wallet you could buy numerous titles from a bookstore, or you could shop online with your victim's credit cards. There is no need to join the rabble of neurotics who enjoy stealing thrillers and episodes of *The Sopranos* from public libraries.

Nevertheless, you can learn much from the more skillful book thieves, who are adept at identifying targets, removing and hiding items, and leaving the scene undetected. Chutzpah is a characteristic of most successful thieves, who use it to their advantage in libraries as much as anywhere else. Following their example, you stroll purposefully into public areas; you do not appear furtive or nervous. You look around for targets, which present themselves in different ways. Many targets are occupied with typical patron activities, such as reading or studying intensely, or working on a laptop, or browsing in the stacks.

Another factor to your advantage in libraries is patrons' undivided attention to their handheld technology. A young woman sits at a table in the reference department and examines the annual reports of local companies. At some point she decides to remove her iPhone and check her e-mail. While doing so, she dons earphones and listens to Beethoven. A message from her boyfriend contains an expression of deep affection. She is enraptured. And that's where you pass behind her and lift her satchel from the back of her chair. Look blasé as you move away. Do not rush. Hide in a washroom cubicle before you examine your prize.

15.2 Deep affection, continued

In many cases, witnesses to your crime will not report it. They might not want to get involved, or they might not understand what they have seen. You looked so ordinary, and moved at a normal pace. Besides, three people who saw you remove that satchel were checking their e-mail at the time, and listening to Bach, or Brahms, or Eric Clapton. Just because they saw you, does not mean that they interpreted your action as wrongful, especially when they were busy composing their own messages of deep affection.

Meanwhile, safe in that washroom cubicle, you have a chance to examine your take. Have a seat. Resist the temptation to rifle noisily through the satchel. Go through it methodically. Remove the wallet and pocket any cash. Then examine the debit and credit cards. These days you will need a PIN to use many of them. See if you can find an appointment book or pocket diary. It might contain lists of PINs, despite the warnings of card issuers against recording PINs in such obvious places.

Look for anything of value that you can sell: jewelry, cameras, and anything gold or silver. If you are not adept at taking advantage of stolen credit cards, you might know people who are. And even if they cannot use the cards owing to the lack of PINs, they might be able to steal the identity of the cardholder, especially if you provide them with other useful items from the cardholder's purse: a driver's license, a provincial identification card, business cards, bank books, a passport, and that appointment book, which contains all sorts of other data that identity thieves can use.

You might also discover a set of keys. If you wish to follow up your initial crime with the theft of the victim's car or burglary of her residence, you must move quickly. But if you are like the majority of library thieves, you will be content with the stolen cash and the credit cards that you can sell to a man who hangs out in a downtown pub and who is involved in the phoney identity racket. He will give you $50 for those cards and buy you a beer.

15.3 Getting out safely

Now you must leave the library without attracting attention. If you have taken only a minute to examine the contents of the satchel, there is a good chance that its owner is still reviewing the loving message from her boyfriend. Tuck the satchel and whatever is left of its contents behind the toilet or in a trash receptacle. If you are concerned that the police will find your fingerprints, wipe down all parts of the satchel that you have touched. In fact it is unlikely that the police will have the time and funding to send the satchel to a laboratory for the discovery and analysis of any fingerprints. And if you tuck the satchel into the depths of the trash receptacle, it might never be recovered.

Take the nearest exit from the library. Again, do not rush. Look normal. Until you are a safe distance away, do not count the money you have just stolen or fiddle with your victim's recently purchased digital camera. (Fortunately she kept the instructions for her new camera in her satchel. You can study them later.) Put distance between yourself and the library, and do not plan to return for at least a fortnight or longer. Tomorrow you can visit another branch, or perhaps the local university library, and find another target.

Whichever library you visit, always pick the softest targets. At the university library, you see a student highlighting his psychology textbook. On the back of his chair is a sports jacket that might contain his wallet and cell phone. But he is young and fit, and does not appear to be captivated by Freud's interpretations of dreams. Lifting that jacket from the back of the chair would be unnecessarily risky. Remember that in larger libraries there are often several targets, and you need not victimize the first that you spot. Look around. Can you spot anything softer? Yes, in that carrel by the window is a well-known physics professor. He has opened his laptop and has been reviewing the doctoral dissertation of his star student, who has compiled fresh and fascinating data on subatomic particles.

15.4 Nature calls

Shortly after you spot the professor, he feels the need to visit the washroom. Still captivated by his student's observations on muons, he walks away from his carrel, leaving behind his laptop, briefcase, and coat, which contains his BlackBerry and the keys to his house, car, office, and laboratory. The keys to the laboratory are clearly labeled as such, against the advice of the university's security managers, who fear that one day a malefactor will gain entry to an area where radioactive materials or hazardous biological agents are stored.

You have about 3 minutes to take advantage of the professor's absence. Stroll over to the carrel and act is if you were the rightful owner of the laptop, briefcase, and other items. Appear to be packing up in preparation to go home. If there are other patrons in the vicinity, they probably will not take a good look at you if they are occupied with their own work. Chutzpah enables you to appear calm and unhurried. Do not leave any of the professor's property behind.

In these circumstances, it is wise to leave the library as soon as possible. You do not want to confront the professor in the washroom if you are carrying items that he can immediately identify as his own. Leave the campus quickly, and examine your new laptop when you arrive home. The professor has made a note of all of his passwords in a notebook that he kept in his coat pocket. You can sell the laptop to a fence tomorrow. The fence will delete all of the data, including the star student's dissertation and the drafts of the professor's next three articles, which are due on his editor's desk next week. (Did the professor and his student back up all of these data? Perhaps. But if they did not, they will be in trouble. Aside from the embarrassment of failing to back up vital data, they might miss deadlines and lose the confidence of their sponsors and superiors. Backing up data is so easy that there is no excuse not to do so. But that is not your concern.)

15.4.1 *In flagrante delicto*

Every thief fears being caught in the act. A patron who sees you stealing his, her, or anyone else's property might angrily demand that you refrain from going any farther. You might hear something such as: "Hey, that's mine. Bring that back," or "Where do you think you're going with that laptop? You stop right now." The worst mistake you can make at this juncture is to run. If you do, the librarian will have reason to dial 911, and the police might catch you in possession of stolen goods. If they don't find you, the librarian will give your description to the police, and distribute it to all branches and even other library systems, thus cramping your style for an extended period, at least in libraries.

The best thing to do when challenged by a victim or any other library patron is to stop and apologize profusely for the misunderstanding. You can offer the most ridiculous excuses, as long as you sound appropriately contrite: "I beg your pardon. That coat [or laptop, or briefcase] looks just like mine, and I've made a mistake." This is usually enough to oblige the intended victim to back down, at which point you might say that one cannot be too careful with so many thieves in the neighborhood. The librarian on duty will agree with you, and all will be well.

15.5 Walkabouts

Be warned that there are a variety of security measures that, if used effectively by libraries, could make theft more difficult. Closed circuit TV (CCTV) and convex (or "fish eye") mirrors to cover isolated areas have been in operation at libraries for decades, but these tools are useless unless library personnel monitor them continuously. In fact, library managers often assume that their installation alone will deter thieves and no further measures are necessary. At many libraries, the most effective antitheft measures include signage that warns patrons about thieves, vigilance, and patrols (or "walkabouts") of public areas.

You should avoid any library that asks staff members to wander around public areas looking for people such as your crooked self. Sometimes these staff members will actually patrol washrooms as well. This sort of security routine is discouraging, but fortunately it is still uncommon. If the majority of library managers organized walkabouts, you could be forced to avoid libraries and seek other sources of revenue. Bank robbery is out, but there are growing opportunities in money laundering, and the future looks bright for manuscript forgers.

And there will be no need to submit a C.V.

16 Into every professional life, a little moon will shine: Dealing with sub-problem patrons

You've been dreaming about it for years. It would feel so good. So do it. Don't hold back any longer. Look that man in the eye and say, "Shut up. For the love of all that's sacred, stop yapping at me!"

That man—a retired accountant—is a regular patron at your branch. He sits in the chair by the reference desk and babbles. You ask him politely if he has a question. He ignores you and starts a monolog about his grade four teacher who made him stand in a corner for throwing a spitball at a boy who eventually became a Member of Parliament, and who had to leave office owing to health problems including gastric reflux and restless leg syndrome, which is especially common in Nova Scotia because nobody speaks Gaelic in Halifax anymore, but they should, and the Babbler will offer you numerous reasons why…

You won't actually tell him to shut up, will you? Of course not. Your library administrators expect you to be courteous at all times. You're obliged to tolerate the Babbler, no matter how irritating he becomes. And he is not the only patron who threatens your sanity.

Another is the Mobile Mum. She pushes a stroller the size of an aircraft carrier through the front entrance to the Children's Department, crushing everything in her path. The stroller contains one child, the day's groceries, a yoga mat, a supply of diapers and toys, a laptop, a purse, a fighter squadron, a company of Marines, and a missile system camouflaged as a series of cup holders. In fact an entire task force of Mobile Mums could arrive at any moment for Storytime. Soon the Children's Department will be crammed with strollers that block aisles and crash into each other, spilling groceries and Marines and children all over the place, and making the Children's Storyteller wonder why the hell she didn't become a pharmacist. But no matter how frustrated she gets, she will remain cheerful and courteous, and welcome the Mums and occasionally the Dads to Storytime. She may not presume to tell parents to dock their carriers outside, but you can be sure that she enjoys fantasies involving the latest model of attack submarines.

16.1 Pass the earplugs

Over the past 20 years, noise levels in libraries have increased. Sources of noise include not only the aforementioned parental navy, but also patrons such as the gentleman in the newspaper section who finds the Sports page dull, and falls asleep. Unfortunately his slumber detonates an arsenal in his pharynx, and you have dubbed him Norm the Snore. His snoring has been measured by seismographs in California, and recently Al Gore suggested that slumbering Norm could provide an alternative form of energy,

Robertson on Library Security and Disaster Planning

if only scientists could develop a method of harnessing the roar that he emits with every exhalation.

Other noise-makers include patrons who conduct long and loud conversations on cell phones, or who simply bellow at each other or you or the circulation staff, as if your library had become mere "public space" such as airport or subway station. You do not dare to tell patrons to pipe down. Gone are the days of the librarian's piercing *Shhhh!*, which current administrators discourage. Gone is the stereotypical librarian who demanded silence and tranquility in all public areas, and who would stifle anyone who dared to make a sound greater than that of a muffled sneeze or a clicking keyboard. Instead you note spiraling decibel levels and see behavior that infuriates you. You recognize that patrons are human beings (most of them, anyway) and that as such they are fallible and inclined to break rules. But you have lost control of your workspace in being forced to allow patrons to act as they please. Now administrators expect extraordinary patience from librarians, technicians, and clerical workers who must deal face-to-face with patrons. Appropriately administrators declare that such library employees are "on the front line," as if they were in a World War I trench. Metaphorically they often are.

16.2 Moon shot

These days, what are the most common outrages to occur in libraries? (Indulge yourself. Consider those matters that administrators call "issues" as *outrages*. Feels good, doesn't it?)

"Clothing styles come and go, and some styles can be problematic," says Cheryl, a thirtyish Adult Services librarian in Vancouver. "For example, the style based on the idea that less is more. Some patrons want to show off their tattoos, and that's fine, but they end up revealing a lot more skin, and some patrons and staff members find that offensive. Some people wear low-slung trousers and no underwear, and when they bend over to take a book from a low shelf, they moon everyone in the vicinity."

In Toronto, a college librarian named Brooke deplores the lack of hygiene among some of the students who appear at her reference desk.

"Most of the kids who attend my college are middle or upper middle class," she says. "They can afford to have a shower every now and then, to brush their teeth and wash their clothes. But some of them don't. And the problem is not gender-specific, either. Young women can smell as bad as young men. Some of these kids have atomic B.O. and horrible breath. I guess there have always been patrons like this, but these days we have to be so politically correct when we talk to them. It's no longer acceptable to tell somebody who reeks to go have a shower. Our administration demands that we respect library users at all times, because to do otherwise would be to humiliate them."

Brooke is often tempted to humiliate one of her smellier patrons, but realizing that she could be reprimanded for it, she remains calm and polite. What might make her snap?

“There’s a creature that uses libraries across the country called the Gum Ghost,” she says. “The Gum Ghost leaves wads of chewing gum stuck to the undersides of tables and chairs. Sometimes he or she simply spits the gum onto the floor or ground for somebody to step on. If I catch the Ghost, I’m going to… Well, I’m going to… ask him or her to deposit that wad of flavorless goo in a trash receptacle. Oh, I’d like to spray the Ghost with disinfectant or air freshener, but that would cost me my job.”

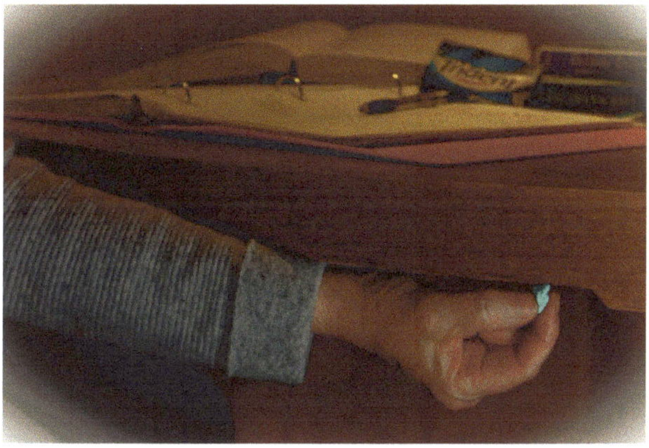

16.3 R & J

The Gum Ghost and foul-smelling persons have patronized libraries for many years. So have Romeo and Juliet, those ardent lovers who engage passionately with each other on the floor in your university library’s Learning Commons, or on an upholstered bench in the Adult Fiction area of your public branch. You would require a crowbar to separate them. If you tell them to desist, they usually will—briefly. And then they will reattach and commence the swapping of saliva, to the horror and disgust of Shouting Edward, who will approach the reference desk and shout at you about the rotten behavior of “those youngsters on the floor over there.” Shouting Edward is a taxpayer, a war vet, a gentleman, and a weapons-grade bore. He will continue shouting until the Babbler arrives and scares him away.

You listen to the Babbler for what feels like an eternity, until you must go and be mind-numbingly polite to the Feeders. These are patrons of any age and gender who insist on treating your library as if it were a cafeteria. They prefer fast food in the form of lukewarm hamburgers and French fries, chocolate bars, gooey doughnuts and muffins, and sugary beverages in paper cups so large that the slurping of cola will create an echo punctuated with the loud clatter of tiny ice cubes. Inevitably the Feeders toss their debris—paper cartons, Styrofoam containers, plastic bags, and cup-and-straw combos—in the direction of trash receptacles, but miss. Trash builds up around the receptacle. Your library starts to look grubby. You ask the Feeders to be sure to deposit their trash in the receptacle, and they mumble willingness to do so through mouthfuls of burger. And they miss that receptacle every time.

"Our administration has posted signage that tells patrons to use trash receptacles, and to clean up after themselves," says John, an academic librarian in the Vancouver area. "Our administrators have given library staff permission to tell Feeders not to make a mess. It would be easier to outlaw eating in our libraries. But now our administration is thinking about setting up a snack counter in our library. Not outside the front entrance, or in the foyer, but right inside the library near the periodical section. This will encourage more feeding, and more mess. We've always had to deal with food stains in books and on DVDs, and I imagine this problem will get worse. But at least the rodents won't starve. We librarians are too kind-hearted."

Meanwhile, as the Babbler babbles and the Feeders feed, the Internet surfers surf, often indefinitely and in all the wrong places. Dedicated surfers show up at opening time and will spend hours at your Internet stations. You may remind them that their time is up, and that other patrons would like to consult online resources. The surfers simply wait until you have turned your attention to something else, and then they move to another station and resume their surfing. Sometimes they investigate porn sites. You have signage that tells them that using library Internet stations to view pornography is unacceptable, but so what? Eventually you lose your temper.

"Turn that rubbish off, you dirty little man. How dare you look at that muck in my library? Do that again, and I will …" What will you do? No, you'd never say such a thing to that dirty little man, as much as you'd like to. Insult him, and he could write to your director and claim that he was examining a medical reference site when you snuck up behind him and called him disrespectful names. He might claim to be an upstanding citizen descended from United Empire Loyalists and many generations of taxpayers, and then you would be in serious trouble. Just as you would be in serious trouble for telling the kleptomaniac to remove those books from under her overcoat before leaving the library. Just as you would be for telling her to stop removing the Tattle-Tape from books in the library's toilet cubicles. Just as you would be for telling the obstreperous drunks and addicts to take their problems to mental health professionals outside your library. You must accept these so-called negative behaviors as part of your job.

16.4 Sub-problem solutions

Judith, a retired library director who lives in Regina, says that she understands the frustration of library employees who must deal with Mobile Mums and Babblers and similar nuisances. She refers to "the sub-problem patron," whom she defines as a library patron who causes problems the solutions to which are usually not worth the effort.

"This is a liberal age, and the library world tends to be generous and forgiving," she says. "Some patrons take advantage of our professional good nature. I hear about patrons refusing to pay overdue fines because they don't feel they should, even though they have the money. That used to annoy me greatly, but I told myself that most patrons paid their fines without complaint. I didn't want to waste time arguing with some selfish neurotic about a small fine, or even a larger fine. I moved on, and I told my staff to do the same. I reminded my staff that sub-problem patrons are not as common as we like to think. The reason that we focus on them is because they grab our attention and irritate us. We let them get under our skin."

Judith says that sometimes it is difficult to tell the difference between a sub-problem patron and a true problem patron, that sometimes the behavioral boundaries blur. She suggests that a true problem patron is one who threatens library employees or other patrons, who could cause significant harm to themselves, other people, or library assets, or who could disrupt library activities "for more than a few minutes."

"If a patron threatens somebody with physical violence, then by all means call the police," she says. "If a thief attempts to steal rare books or artwork, then dial 911. But if somebody with emotional problems tries to walk out at the door with a battered romance novel or two under her coat, don't bother raising the alarm. The police have far more serious crimes to investigate."

As for the other sub-problem patrons, Judith recommends "firm assertiveness." You may want to scream at the Babbler, the Feeders, and the Surfers, and you may long to spray Romeo and Juliet with a fire hose. But a better and more acceptable approach is to express your disapproval in polite but nonetheless vigorous terms. Establish strong eye contact. Speak plainly and clearly. Ensure that your posture is healthy and upright, since poor posture will make you seem smaller and less robust. Repeat yourself as required, and be prepared to repeat yourself a number of times. Do not confuse assertiveness with bad manners.

Develop a presence in your workplace. Practice constructive disruption of inappropriate conduct. In time, you will earn a reputation for being in charge—which you will be. It's a matter of reclaiming your workplace from those who would take it from you. When you reclaim it, not even a task force can stand in your way.

Wise to the forger's tricks: Detecting bogus signatures and inscriptions in collectable books

17

Warning: The first edition that you were about to buy from an antiquarian bookseller contains a phoney signature. Like most attempts at forgery, it is clumsy and obvious to anyone familiar with the author's genuine handwriting.

Look closely at the ink and the nib width. The forger has used a ballpoint pen to fake Charles Dickens's handwriting. Dickens did not live in the Age of Bic, nor was he familiar with the style of handwriting taught in North American elementary schools during the 1950s. But the bookseller has accepted that signature as genuine, and wants several times the usual price for the book. You express misgivings about the signature's authenticity. The bookseller swears that it is Dickens's. Not wanting to spoil your relationship with him, you politely refuse the book and change the subject.

A month later, you spot an entry for the book in another bookseller's catalog. The price is even higher than before. The description mentions "a superb signature on the flyleaf." Curious, you telephone the bookseller and confirm your suspicions. You also learn that she has just sold the book to the special collections department of a university library. Evidently the librarian in charge of the purchase is a Victorian scholar, and thrilled to obtain a signed Dickens for his collection.

Of course, knowledge of Dickens's characters and syntax does not translate into expertise on his handwriting. Hence many well-educated booksellers and buyers fall for forged signatures and inscriptions. If the forgery is exposed, the matter is often hushed up. In some cases, the owner will continue to treat the book as if the author's pen had actually graced it. The truth is sometimes too embarrassing to acknowledge.

Forgery is as old as pictographic writing. It took only one crooked scribe to understand the advantages of faking a mark or seal on a clay tablet or papyrus; thereafter the crime spread wherever literate cultures flourished. Most forgeries occur during financial transactions—on wills, property titles, and correspondence related to trades and other deals. Occasionally the media report a spectacular forgery in the art market, a bogus Rembrandt or Picasso, or a piece of sculpture from a suspicious workshop. Literary forgeries hit the headlines less frequently. Reports of phoney signatures on flyleaves appear rarely, hidden away in obscure collectors' newsletters and journals. It is reasonable to assume that literary signature forgery is an unpopular topic among those who derive pleasure and income from collectable books.

Over the past decade, however, as stock markets fluctuate, interest on deposits decreases, and investors look desperately for safer investments, one hears more about

Robertson on Library Security and Disaster Planning

suspected frauds on bookshop shelves. Criminals have discovered that they can buy an early edition of Robert Frost or Wallace Stevens for under $50. And then with less than an hour's research online or in a local library, they can scribble a passable signature on the flyleaf and title page and sell the book for $500 or more. The buyer of such an item could be a private collector, a bookseller, or a library. Investors plunge into the shark pool of eBay and surface with the crudest attempts at signature fraud.

Some years ago in Toronto, a bond salesman discovered 40 volumes of Beat poetry for sale on eBay. Each was signed on the title page by Lawrence Ferlinghetti, or Allen Ginsberg, or Gregory Corso. A closer examination of the signatures revealed that the forger made no attempt to copy the poets' genuine signatures. In fact, the signatures had been clearly written by the same hand, with no difference in vowel shapes, loops, or spacing between letters. The same ink appears on every title page, evidently from the same pen. The salesman was delighted to pay just over $15,000 for the collection. The real value, based on the availability and condition of each volume, was under $500. As in other investment strategies, bonds would appear to be the safer investment, although the salesman suggests that someday his Beat books could appreciate in value as forgeries. He might be right. Forgery collectors exist, although they usually prefer more deft attempts at signature duplication.

Forgers are especially adept at sending their products across borders. American and British buyers have bought works by Margaret Laurence that she signed years after her death—about which the buyers were unaware. Margaret Atwood remains above ground, and, according to European dealers, she spends much of her time scribbling inscriptions to men identified only as Helmut, Jean-Claude, and Franz. Either she is terribly fond of Franz, and has inscribed dozens of copies of *The Blind Assassin* to him, or she knows and respects multiple Franzes and wants them all to own her Booker-winner. The mystery might never be solved, even by Atwood.

Some forgers are highly skilled. In fact, librarians and booksellers who know how to detect a bogus signature or inscription admit that many of the better forgeries will never be brought to light. An artful forger's work is convincing not only because of the amount of time he or she has dedicated to mastering an author's signature style, but also because he or she has succeeded in imitating that author's psychodynamics. Essentially, the forger must develop an ability to mimic the mechanics of jotting down the author's signature and, at a deeper level, imitate and implement the author's mental state as it would function during the act of holding a pen and pressing it onto paper. Without the right psychodynamics, an act of forgery will not deceive many experts, or even many knowledgeable amateurs.

Sadly, however, there are too many prospective buyers for what are sometimes called "bad signings" or "fraud copies," no matter how patently false they are. Booksellers wince when they describe a collector's reaction to being told that a prized first edition of Henry James or Edith Wharton has been inscribed with a felt pen less than a year ago. Sometimes the collector refuses to accept the bookseller's verdict. The collector might then visit a rare book librarian and ask for signature authentication. Even with a sample of James's or Wharton's real signature in plain view, the collector might still refuse to admit the truth. But the book might appear shortly thereafter on eBay, or on a bookseller's website, or at an auction, and the fraud copy circulates until it finds a more permanent place on a shelf in a library or collector's study.

By the way, you'd be wise to examine your other signed Dickens editions. Do you know for sure who held the pen that signed those flyleaves?

17.1 An author's inscription: What to look for

Before you buy the book:

- Compare the author's handwriting on the title page or flyleaf to an authentic sample, if available. A person's handwriting—and especially his or her signature—changes over time. Use a sample that comes from approximately the same period of the author's life as the signature or inscription in question.
- Look for signs of hesitation within particular words. If a forger has not developed the appropriate psychodynamics, you will find breaks between letters, and sometimes breaks within individual letters—evidence that the forger has lifted the pen in an attempt to copy a letter's correct shape.
- Beware of anachronisms. Victorian authors did not use felt pens. John Updike did not use a quill, and did not produce copperplate at book launches. Nor did Hemingway use a purple rollerball.
- Beware of inappropriate locutions. Tennyson would never inscribe, "This one's for Ed" in a copy of *In Memoriam* intended for Edward Fitzgerald.
- If it seems too good to be true, it's probably bogus. Remember the Hitler diaries.

18 Know thy vendor: Getting the best in off-site records storage

Librarians and other information professionals should be careful about where they store their records. Some off-site vendors are better—and more honest—than others. Much depends on reviews of a vendor's site, reputation, and indexing system.

I did 6 years in a Federal penitentiary that reminds me a lot of the warehouse I run. It's an old concrete pile and it smells damp. You know the place. You store your records there: dozens of boxes crammed with files and miscellaneous materials. Much of this stuff is junk. You'll never need to look at these old correspondence files again, or those cartons of brochures entitled *Hot New Fiction for 1983* and *Welcome to the World of DOS*.

Why are you saving so much obsolete stationery? How about these boxes of letterhead with the logo that your library stopped using two decades ago? Where did these antique computer tapes come from? And what's so important about these moldering newspapers and telephone directories? You may not remember these items because so much time has passed since you packed them up and sent them my way. The only reminder I give you is a monthly invoice, which you pay without question. I love this business.

But not everything that you store off-site is worthless. For example, consider these original building plans for your Central Library and all of your branches: not only the structures, but also the plumbing, electrical, and heating systems. This material is essential for the purposes of facility maintenance. It even includes warranties for the roofs and other building components. You wouldn't want to lose these documents, would you? If you did, it would cause a lot of inconvenience for any contractor employed to work on your buildings. The cost of the work could increase a great deal. And without the warranties for the roofs, you would have to pay far more for repairs.

18.1 City hall

"What people tell me is that they're not worried about their building plans because somebody else has copies of them," says Brian, a construction contractor with 30 years of experience across North America. "Departments connected to a municipality assume that city hall keeps vital records and is prepared to share them at short notice. But often those records are lost, or misplaced, or simply unavailable. I'm obliged to work without them, and my company has to spend more time—and more of the customer's money—figuring out what's in the plans so that we can proceed with our project. I'd advise libraries not to take the storage of their records for granted."

Your trustees will suggest that your valuable files need better storage. You ran out of space in your branch basements years ago, so you need an off-site storage vendor who will treat your information with respect. But first, what about your records management?

If you want to control costs by cutting down on useless paper, you'll need an inventory of the records in all of your locations. Your inventory should contain the most basic data regarding the kinds of documents you keep, their inclusive dates, quantities, and current storage location. If you compile an inventory from scratch, you'll probably discover tons of material that you don't really need. Your inventory can turn out to be an effective cost-saving tool, since it will indicate opportunities to free up space and reduce storage expenses.

Next, you need a records retention schedule, which tells you how long you must retain various documents. Does your accounting department already have a retention schedule? That's good, but is it up to date? Does it indicate retention periods for all of your library's documents? You must audit and update your retention schedule regularly so that documents from all departments are controlled. Without a well-maintained retention schedule, you'll pay a lot for the storage of paper that should have been sent to a recycler.

18.2 Procrastination

"People often neglect to keep their retention schedules up to date," says Lauren, a records manager and corporate librarian in Toronto. "There's simply too much to do, and some tasks are more urgent than others, and the retention schedule can be ignored for another week, or month, or year. And eventually it's so out of date that you must spend a lot of time updating it. But without it, managing your records—particularly in off-site locations—can be more difficult and a lot more expensive."

When you decide to store records off-site, make sure that your off-site vendor is reputable. It's amazing how often organizations in all sectors send sensitive information to vendors with questionable backgrounds. You should know better than the Attorney General's office that stored sensitive records concerning criminal investigations in a warehouse owned by mobsters. Or the trust company that kept its mortgage files in the basement of an office tower managed by an offshore company that specialized in shady property development. Or the insurance company that sent confidential medical files to a paper destruction company that sold the files to a movie production company.

You should know better, but do you? Have you actually checked out the reputation of your off-site vendor, asked for references, interviewed the warehouse manager, and made inquiries about the vendor's reliability? If not, you might be keeping people like me in business. If your confidential files are used for unscrupulous purposes, you might never find out. Worse, the illicit use of your records might be exposed, and you'll have a crisis to deal with. Your trustees will not like the idea of your sensitive information falling into the wrong hands. How would you and your coworkers like the idea of your HR files falling into the wrong hands? You might stop doing business with me, but I won't suffer much. I may deal with a lot of paper, but I never leave a paper trail. Moreover there are always new customers for a vendor who has storage space for paper.

18.3 Pigeons unacceptable

You should also inspect any off-site space in which you intend to store records. Be sure that the warehouse is well maintained. Look for signs of water ingress and moisture damage. Check the security procedures and fire control systems. And examine the shelving: is it stable, moored to a supporting surface, and rust-free? Do you see any sign of rodents or pigeons? Don't put up with sloppy warehouse maintenance or substandard security, and don't accept a vendor's promises that a warehouse's weaknesses will be rectified soon. That vendor might assume that you intend to visit the warehouse only once. Usually he's right, since many customers don't even make an initial visit.

An off-site warehouse may be in top condition, with state-of-the-art security, but what about the vehicles that transport your documents to and from your sites? Some customers prefer unmarked vans with alarms and advanced locking systems. Others will tolerate any old truck that pulls up to the loading bay. Do crooks in your area break into vehicles? Do they steal anything on wheels? If so, ask your vendor to use an inconspicuous and fully alarmed vehicle when transporting your files.

"There are numerous cases in which boxes of documents go missing in transit," says Drew, a records manager in Western Canada. "I've heard about boxes flying out through the back doors of vans on the way to storage warehouses. The boxes hit the highway and paper flies everywhere and spreads for miles. That is a literal paper trail as well as a records manager's nightmare."

To ensure that your files aren't misplaced in an off-site location, check the vendor's indexing system. A responsible vendor won't accept your materials for storage unless they are clearly labeled with your library name and other identifiers. Large vendors use bar coding to track each document box. Ideally the vendor will be able to tell you exactly where your records are at any time. Some customers use their vendor's indexing system to compile a records inventory, and work with the vendor to develop retention schedules.

18.4 Additional services

Once you find a vendor whom you can trust, consider the ways that you can work together to manage your company's records. Vendors can often provide a lot more than storage space for documents. Some will destroy and recycle old paper, or arrange for its conversion to digital formats. Some will assist you with your business continuity planning, and set up hot-sites and alternative space in case your administration offices are damaged or inaccessible. Many offer records management auditing services to discover any inefficiencies or security weaknesses.

Large vendors maintain data storage systems for clients who need to back up their electronic information off-site. Beginning in the 1970s, customers would store their tapes and disks in the vendor's vaults, and some still do. But the current trend is to transmit data electronically to an off-site system in a secure location. That location may be in the same city as the client, or across the country, or on another continent. These days, collocation technology and the Internet allow clients to have multiple backup sites at a reasonable price.

Off-site storage of electronic data demands comprehensive security measures. Your firewalls and encryption must be current, high quality, and frequently audited if you are to avoid losing data to hackers. Your off-site data security systems should be strictly confidential, and known to only your senior IT managers. Materials related to security—for example, manuals—should be carefully locked away. You can be sure that hackers would enjoy browsing through them.

18.5 Contract or con man

Whatever you demand from an off-site vendor, include your terms in a contract. If you're a new client, feel free to sign a contract for no more than one year, to see how well your vendor performs. Don't agree to a longer period until you're satisfied with the vendor's service, facility maintenance standards, and prices. Insist on a cancelation clause. Above all, don't compromise your library's standards just because a vendor won't agree to improving security measures at a warehouse or testing an electronic backup system. Remember that you and your vendor should work together so that you can derive the greatest benefit from your off-site storage agreement.

Or you can take your chances with a vendor like me. You've never visited my warehouse or asked me about its security systems. You don't care about the water stains on the ceiling, or the rats, or the occasional loss of a document box. It doesn't matter if a few records go missing, especially if they're garbage. But sometimes they're worth their weight in gold. Look at these borrower records you sent me accidentally. I've got a friend who puts together mailing lists—but you don't want to hear about that.

Note: The names of persons quoted in this chapter have been changed to ensure privacy. And no, Mr. Robertson has not spent time in a Federal penitentiary. He remains at large.

Healthy snacks, safe stacks: How one library enhanced its security 19

In virtuous British Columbia, a library's occupational health and safety (OHS) committee has replaced the coffee and doughnuts at its meetings with organic celery and carrot sticks, which committee members wash down with the purest bottled water. The result of this change in snacks is remarkable. OHS meetings used to fill two hours. Now the committee reaches agreement on most points within 45 minutes.

"There is a positive correlation between healthy food and reduced meeting duration," says Glen, the committee's chair. "That's probably because people are sick of celery and carrots, and everybody wants to dash down the street to the coffee shop that sells gooey cinnamon buns and cupcakes. Anyway, the committee has made some interesting decisions lately regarding security at our main library and branches."

The library's director asked the OHS committee to identify security problems that might arise over the next decade, and to recommend ways to solve them. What Glen and his colleagues determined would interest librarians worldwide.

19.1 Anything can go

It is not surprising that theft of library materials remains a risk at all of the library's sites. What has changed is the variety of items that might disappear. In the past, thieves stole popular books and magazines. Now they attempt to steal anything that they can carry, from any book, magazine, CD, or DVD to keyboards, chairs, and pencils and pens from circulation counters.

"Nothing is safe," says Glen. "We've had posters removed from the walls of a children's department. At the same branch, the umbrella stand by the front entrance went missing. At another branch, somebody tried to break into the public photocopier to steal the coins. And there have been a number of incidents involving unauthorized persons entering staff areas. On one occasion, a librarian had her purse stolen from her locker. The thief also went through staff members' coats in the cloakroom and stole two sets of car keys."

Fortunately the key thief did not succeed in stealing anyone's vehicle. Glen assumes that the branch's parking lot was too busy, and that the thief did not want to attract attention.

"Nevertheless, the thief had a lot of gall," says Glen. "He or she was also quick and clever enough to get into our staff area unnoticed."

After this incident, Glen and his committee members had to consider several uncomfortable questions. First, could a staff member have committed the theft? Second, if anyone had caught the thief in the staff area and in the act, might there have been a risk of violence? Third, might the thief strike again? And finally, would the security measures in place at the branch be effective in preventing similar incidents?

"We investigated the matter thoroughly, as did the police," says Glen. "We're satisfied that an outsider committed the theft. We also believe most thieves are not likely to become violent, and that if they were caught in the act they would try to leave the scene as quickly as possible. As for a repeat performance, another theft, that's always a possibility. And even with an advanced security system—for example, live-monitored CCTV—our library would not be entirely secure, since no security system can guarantee absolute security."

19.2 Temper, temper

Aside from theft, Glen and his colleagues considered incidents involving angry patrons who verbally threatened staff, who mishandled hardware such as keyboards and photocopiers, who vandalized print materials, and who damaged furniture and washroom fixtures.

"We had to distinguish between wear-and-tear and willful damage," says Glen. "With the increasing use of different branches, we'll see a lot more wear-and-tear, which we consider the cost of doing business. But we're also seeing an increase in willful damage. Fortunately it will not pose serious risks to staff or patrons, but library management must take it into account at budget time. Damage costs money."

Inevitably, online misbehavior and crime became a topic of discussion. The usual problems came under scrutiny: accessing pornography at public terminals, hacking, electronic bullying, and vandalism. Monica, the library's systems librarian, suggested that while there were security measures to discourage such activities, there would always be a risk, and the library would never be able to prevent all attempts to access pornography sites or send abusive messages through library terminals. Nor would it be possible to detect even a fraction of such incidents.

"Anarchy rules on the Internet," says Monica. "There will always be ways to work around security filters. We have to face the fact that porn users and hackers can be as clever as the programmers who create the filters and related protocols. This does not mean that we should give up our attempts to discourage ugly behavior, but we should not expect high levels of success in preventing it. That's the realistic approach."

19.3 Meeting set-up

Glen organized a series of meetings for the committee to decide upon its recommendations. He set the goal: the formulation of a list of the library's ongoing and future security risks, with a series of clearly stated recommendations to deal with them. He then scheduled three meetings—complete with organic vegetables—to finish the job.

"I've been a librarian long enough to recognize the possibility of never-ending planning processes," says Glen. "It took another committee months to make a decision about replacing the furniture at one of our branches. In the end, the furniture started falling apart, so the decision virtually made itself. I didn't want to spend months arguing about anything, so I limited our discussion time and told committee members that there would be no postponements or delays. This approach worked well."

The committee decided that current security problems would continue to exist in future, with increasing frequency at branches in growing neighborhoods. At some branches, there could be more attempts by thieves to steal patrons' portable electronic devices such as laptops and cell phones. One hopeful note: the risk of violence toward staff might not necessarily increase over the next decade, and at some library sites it could decrease. To address all of these circumstances, the committee made a series of recommendations that the library's management and board have approved.

First, in future the library's security will be reviewed on its own, and not as an aspect of management in general, OHS, or disaster preparedness. Considering security matters alongside other topics can blur the focus and lead to confusion. Thus security plans will not appear in binders or on the website with earthquake preparedness plans and first aid tips. Moreover the committee will dedicate certain meetings to security matters only. In the words of the library's director, security will have a "fresh, independent status" at the library.

Second, the committee will compile a security report for the entire library every 6 months. This report will be based on incident reports from all library sites as well as staff members' observations and anecdotal accounts of security-related issues as they arise. In the past, incident reports were noted at only a few of the library's sites. There was no standard incident report form or folder for each site. There will be from now on, and the department manager or branch head will be responsible for maintaining these records.

Security reporting will include not only descriptions of actual incidents, but also warning signs of events that might happen. Examples include: the deteriorating behavior of a patron, who seems to have serious emotional problems and who could threaten staff or other patrons; damage to locks on doors and windows that indicate attempted break-ins; scurrilous and threatening graffiti; the discovery of weapons including knives and firearms; and suspected stalking of staff or patrons. The reporting of these warning signs will alert staff to possible security breaches, and encourage vigilance in specific areas of the library and at certain times. Attention to warning signs might, in many cases, prevent actual incidents.

"We want to enhance our workplace violence policy," says Glen. "That is, we want to make it more than just another binder and training package. We hope that staff will learn to recognize the warning signs of security problems that might involve violence and physical danger. Forewarned is forearmed, as the cliché has it. We want that cliché to make sense to every staff member."

19.4 Orientation and training

Third, the library will spend more money and time on security orientation and training, and all staff—including part-time and auxiliary employees—must participate. New hires will receive an orientation package on their first working day, and the Human Resources manager will make sure that they read and understand the material.

"Realistically, we don't expect everyone to become security experts by reading their orientation package," says Beverly, the library's HR manager. "The package is basic information on staying secure while working at the library. There is advice about inappropriate patron behavior, bullying, tampering with equipment, and vandalism. There is advice in bold type regarding 911 calls and when to make them. There are also directions about reporting these things immediately to one's manager and making the necessary report. But will people take the orientation package seriously? Probably not. So we need to reinforce the initial security advice with follow-up training."

In the past, the library has invited a security consultant to attend meetings for branch managers and department heads and to discuss general security matters. Now, the library will invite the local police to give brief (i.e., half-hour) talks on topics covering different kinds of warning signs, relations with potentially violent patrons, and dealing with thieves and vandals. Training will also include lunch-and-learn sessions for all staff with local workplace violence consultants, newsletter discussions of internet security risks, and—when the budget allows it—the development of a webinar regarding library security. The committee hopes that the latter tool can eventually replace the orientation package, and that it will be easier and less expensive to update.

Fourth, the library will install more security-related signage at all sites. This will include signs that advise patrons to treat circulation counter staff with respect, and that inappropriate behavior will not be tolerated. There will be signs in all public washrooms regarding regular patrols by library staff; there will also be signs discouraging washroom graffiti.

"We're not sure how effective the antigraffiti signs will be at some branches, but we're willing to give them a try," says Beverly. "If a sign persuades a single graffiti artist not to carve an offensive drawing into the paint on a washroom cubicle wall, then the sign will be worth the cost. We have spent a lot of money removing nasty graffiti."

19.5 Exterior risks

Another addition to the library facilities will be CCTV cameras and better lighting in parking areas where there have been car thefts and break-ins, and in one case a mugging. While the library has avoided hiring security guards in the past, now it will establish a contract with a security firm for patrols of the exteriors of branches in high-risk areas. At present, security guards are not deemed necessary inside those branches, but if crime increases in their neighborhoods, guards could be asked to patrol inside entrances, public areas, and washrooms.

"We are of mixed minds regarding patrols by librarians and clerical supervisors," says Glen. "Some branch heads accept this responsibility without hesitation; some have been patrolling their facilities for years. Others don't like the idea, because they fear what they might come across in secluded areas of the stacks or in washrooms. What we might do is to canvass staff at every site and find out who is willing to make patrols. We would make the task voluntary. I believe that there are staff at every site who would be willing to do it."

An overarching matter that the committee will consider regularly is the problem of apathy with regard to security. In many organizations people take it for granted, and eventually measures to safeguard staff and assets are forgotten—that is, until a security breach occurs and managers are forced to deal with the consequences.

"We intend to avoid apathy and keep our security measures up to date and effective," says Glen. "It doesn't require an enormous amount of effort or a big budget to do so, even with the increasing risk of breaches. After all, it's not as if libraries are frequent crime scenes. Our brief, regular OHS committee meetings have served us well, and it's fair to assume that our library is one of the most secure of its kind in the country. I urge others to follow our example."

Celery, anyone?

Section B

Disaster planning

Hoping for the best, preparing for the worst: A disaster planner's experience

20

Library history contains numerous accounts of disasters, from the destruction of the Alexandrian Library in the third and fourth centuries AD to the loss of millions of volumes throughout Europe during World War II. Every year, dozens of libraries lose part or all of their collections owing to fires and floods. These risks have prevailed since ancient times, along with earthquakes, severe storms, war damage, and sabotage. There is no reason to believe that the world will become any safer for libraries in future. In fact, with our increasing dependence on electronic media that are even more vulnerable than printed materials, it is possible that libraries will sustain serious losses more frequently.

None of this will surprise scholars and librarians who understand that the study of any corpus of writings is essentially the examination of what has survived over time and spite of dire circumstances. For example, when we take a course on Sophocles (496–406 BC), we are able to read only seven of his tragedies and a few miscellaneous fragments of his other works—a small part of an oeuvre that once comprised more than 120 plays. What has happened to the rest? Like most classical literature, it has vanished during one calamity or another at some point in the distant past. In light of the long list of risks that manuscripts are prey to, it is remarkable that any ancient literature has survived at all.

While nothing lasts forever and the eventual loss of library materials is unavoidable, at least there are methods to retard the process of decay and allow collections a longer lifespan. My career as a disaster planner is the direct result of my fascination for those methods that enable us to preserve our vital information and cultural heritage. To put it simply, as a library disaster planner, I am an idealist with an interest in hazardous plumbing.

I came to librarianship with a background in anthropology that included studies in archeology and Egyptology. At one time I considered a career as an Egyptologist, but was deterred by the thought of the enormous commitment necessary for what might have been exceedingly small returns. After all, there are very few jobs for specialists in hieroglyphics, and I have yet to see a classified ad for an expert in antique funerary practices. But to this day, I am delighted to read anything concerning Egyptian writing and literature, and am struck by the irony of the fact that the texts carved into the walls of the pyramids have a better chance of surviving longer in physical form and human memory than the books listed in my local newspaper's bestseller list.

Something else that lured me away from Ancient Egypt was the study of rare books and bibliography. When I learned that I could devote much of my time to these and related topics at the University of British Columbia School of Librarianship (as it was known in those days), I sent in an application. Two years later, I graduated with the

Robertson on Library Security and Disaster Planning

usual degree and the usual crushing debt from student loans. Here luck intervened, and I found two part-time jobs that gave me an opportunity to learn the rudiments of disaster planning, as well as to earn a living.

First, I found myself in charge of the library at Thurber Engineering Ltd. in Vancouver, BC. Thurber specializes in the investigation of natural hazards, such as earthquakes, landslides, and avalanches. In the late 1980s, the company formed a subsidiary that deals with environmental hazards, including toxic spills and marine pollution. Among my duties were the development and updation of the collection to reflect advances in all areas of hazard research. As would be true for any librarian in my position, I became familiar with the terminology and general subject areas of hazard evaluation and management. Above all, I became aware of various methods of risk assessment which, in slightly modified forms, I use in my current projects.

Aside from my job at Thurber, I took on the role of consulting librarian, archivist, and records manager at Seaboard Life Insurance, which had offices across Canada. My main job was to devise survival strategies for Seaboard's archival collection and vital records. Eventually I became involved in the company's microfilm conversion program, an ambitious project that, when completed, guaranteed the long-term survival of millions of important documents. Each individual paper document was flattened, unstapled, filmed, indexed online, and finally recycled. In the event that Seaboard's head office in Vancouver is destroyed, its vital records will survive in microfilm stored in secure off-site vaults. Using the microfilm collections, Seaboard staff will be able to resume basic business operations within 72 hours.

In 1985, with a moderate amount of experience and technical knowledge, I started to work on disaster planning and security projects for special libraries. These projects usually included the comprehensive backup of databases and the safe storage of valuable materials. Special librarians were also concerned about the enhancement of corporate confidentiality, especially with regard to new products, marketing campaigns, and personnel records. They were interested in finding new ways to reduce what is politely referred to as "internal misplacement" and which is more accurately described as "staff theft".

At that time, my practice as a library disaster planner was limited to the West Coast, although occasionally I found work in eastern Canada and the United States. As my client roster grew, so did my interest in areas such as emergency management, disaster response, and business resumption. I earned a first aid ticket and attended seminars on conservation, computer fraud, alarm technology, white-collar crime, and crisis management. I must admit that much of my professional development was more theoretical than practical, but I saw no reason to change my ways. I was in the perfect frame of mind for the proverbial shake-up, which came to me just as I was reaching for a paperback copy of Albert Camus's *The Plague*.

I was attending a conference in San Jose, California. The topic of the conference was bar coding. I have great respect for bar coding; I believe that in many settings it is an excellent way to enhance security while increasing efficiency. But several days of presentations on bar coding is not an exciting prospect. In fact, several minutes of bar coding is all that most of us can bear.

Just before 5 p.m. on October 17, 1989, with enormous relief, I walked out of the San Jose Conference Center into a warm evening. It was a World Series game night, and the streets were mostly deserted. A sign outside a mall announced that its bookshop remained open for business. I was looking for reading material aside from brochures on code scanning software, so I popped in for a browse. I was not impressed with the palette-loads of Stephen King and Danielle Steele, and there was nothing for me in the magazine rack, which contained every surfing magazine published in the western hemisphere. At the back of the shop I found a shelving unit crammed with what a sign described as "Lit'n'Stuff." This sounded considerably more promising than *Surfin' Tan Monthly*, but I was disappointed to find little more than a row of gift-edition Shakespeares and *The Collected Poems of Rod McKuen*. Then, to one side, I spotted a copy of Camus's *The Plague*, which looked like the sort of thing that could clear my mind of bar coding. I reached for it. And then, all of a sudden, the shelf seemed to reach for me.

At least it hopped toward me, McKuen and all. The lights dimmed, then went out. All that anyone could hear at the moment was a deep, grumbling roar. The building began to shake, the floor undulated. The shelf hopped again in my direction, and showered me with Shakespeare. And then the shaking stopped, and everything fell silent.

My trip to California turned out to be more educational than I had expected. I may have forgotten everything I learned about bar coding, but the Loma Prieta earthquake was the most compelling teacher that any disaster planner could ask for.

After the quake, I spent several days inspecting various sites that had been damaged, including a number of libraries. Certain images are indelible in my memory: walls with deep cracks from floor to ceiling, aisles buried under thousands of fallen volumes, row after row of collapsed shelving, piles of shattered glass. I soon realized that many of my notions regarding disaster preparedness and response were impractical. I also understood that, while most Californians are fatalistic about many risks that prevail in their state, my fellow Canadians will ignore such things until it starts raining Shakespeare.

Upon my return home, I revised my disaster planning methodology. I developed a commonsensical model in the form of a schedule or action plan that my library clients could follow to mitigate the risks that prevail at their sites. I made sure that the model could be adapted to meet the needs of librarians in different kinds of organizations. What I changed more than anything else was my approach to library projects and clients. Instead of assuming responsibility for all aspects of library's disaster plan, I changed my role to that of trainer and facilitator for in-house disaster planning activities.

Currently, my relationship with most library clients begins with a call from a librarian who has been assigned the task of "putting together some kind of plan in case the roof falls in." This librarian usually has no experience in disaster planning, but he or she has time to spare—or so the occupational health and safety committee assumes. Where to begin? Nobody on staff seems to know. During our first meeting, I explain five essential points of disaster planning.

First, the best motivation for compiling a disaster plan is a controlled sense of urgency. In other words, what should motivate you is not panic, or boredom, or a little bit of money that has to be spent before the end of the year, but an awareness that there are risks to your library's safety and security. You know you should take effective steps to deal with those risks. There is no such thing as a risk that cannot be mitigated.

Second, effective planning is site-specific. This means you must pay strict attention to your own library site or sites, and resist the temptation to borrow the plan of a library located in another region. If you work in a public library in Winnipeg, Manitoba, then you should concentrate on the risks that prevail in Winnipeg. The disaster plan compiled for a college library in southern California will not give you the background you need to prepare your library for its prevailing risks. After all, Winnipeg suffers very few earthquakes, and Southern California is not known for its snowstorms.

Third, you should depend on facts, not on uninformed opinions, guesses, or gossip. There are often false rumors about risks that prevail at certain sites. These rumors can lead to bad planning. For example, librarians in Alberta and Saskatchewan need not worry about earthquakes, yet I have seen several disaster plans for libraries in these provinces that include procedures for staff to follow when the shelves start hopping about. Apparently, some years ago, a rumor circulated that there was a serious earthquake risk across the Prairies. This is not so. Ironically, in none of these plans were there effective measures for dealing with flooding, which is the most common form of disaster in libraries across Canada.

Fourth, your model for an effective disaster plan should address the needs of your library before, during, and after any disaster. Accordingly, your plan will include three sections: preparedness, response, and resumption.

Fifth, the effectiveness of your disaster plan is ultimately up to you. A consultant can give you helpful advice on anything from evacuation procedures to data backup, but in the end, you and your colleagues are responsible for the major components of your plan. It is up to you to test the plan and keep it current. No consultant can promise to be onsite with you when all hell breaks loose.

In subsequent meetings, my contact and I can begin working on the library's disaster plan. In most cases, I set a time limit on the planning process so that it does not become too lengthy and expensive. Generally speaking, I like to work with my library client to produce the first draft of a comprehensive plan within one month or less.

We begin with the preparedness section. This will contain a summary of risks to the library, including any outstanding natural risks, such as earthquakes, severe storms, high winds, and floods from bodies of water, such as rivers and lakes. Risks arising from human activity should be noted as well. These include toxic spills, sabotage, floods from faulty plumbing and roof leaks, and fires from faulty wiring.

At this point in the planning process, it can be helpful to distinguish between an emergency and a disaster. An emergency is an event that can be managed and resolved either by traditional "first responders"—police, firefighters, and ambulance crews—or by a person or persons at the scene who can initiate emergency response procedures. For example, you are faced with an emergency when a senior has a heart attack in your reference section. There should be at least one staff member onsite during opening hours who can administer CPR, and all staff should know when it is appropriate to summon an emergency team by dialing 911. Emergencies are usually limited to one site. A disaster is an event that is beyond fast resolution owing to its size and severity. For example, a toxic spill can force the evacuation of an entire community and can cause multiple casualties. Such an event cannot be easily controlled by any human agency, and often recovery from it will be expensive and time consuming. An effective disaster plan will protect you in the event of either an emergency or a disaster. In preparing for the big one, you should cover the small ones as well.

Having identified the risks, we try to find ways to mitigate them. If there is any sign that the plumbing in the library is inclined to leak or burst, we arrange for it to be inspected. If there are concerns about any sort of fire hazard, we invite the local fire department to visit the library for a tour. In earthquake zones such as British Columbia, we will recommend that all shelving be moored securely to the floor. We will find out which staff members have training in first aid and CPR, and encourage the rest to take courses from the Red Cross or St. John's Ambulance.

The preparedness section also contains a record of all fire drills, when they took place and how well they performed in evacuating the building. Was the entire staff outside in less than a minute, or did people dawdle? To some, drills may seem to be a waste of time, but they should be reminded that in a real fire, a cynical attitude could cost them their lives.

Our next step is the formulation of the response section, which addresses the risks indicated in the preparedness section. It should be brief enough to fill no more than a single page, which can be easily reproduced and distributed to every staff member. Usually in point-form, it will cover measures that everyone should take to protect themselves during an actual disaster or emergency. The following response section is part of a plan in effect at a Vancouver special library.

20.1 Emergency/disaster response—Please read this now

20.1.1 In the event of a fire or explosion

- Avoid flames and fumes.
- Remember: the closer to the floor, the lighter the smoke and fumes.
- Small fires may be doused with fire extinguishers. In the event of a spreading blaze, trigger the fire alarm and evacuate the premises immediately.
- Allow no one to remain behind.
- Once outside, call 911 to ensure that the fire department has been contacted.

20.1.2 In the event of an earthquake

During the shaking

- Take cover under sturdy furniture. Protect your head.
- If indoors, stay indoors; if outside, stay outside.
- Avoid glass, loose masonry, and utility wires.

After the shaking

- Apply first aid if necessary.
- Do not use candles, matches, or open flames.
- Do not smoke.
- Do not make telephone calls unless they are life saving.
- Turn on radio or TV for emergency bulletins.
- Avoid entering damaged buildings. If in doubt, stay out.
- Expect aftershocks.

20.1.3 In the event of a flood—Most floods are not life threatening, but

- avoid wet wires and electrical outlets
- turn off water mains if possible
- do not begin clean-up until after an inspection of electrical outlets.

This response section covers risks for one library site only. Different sites in the same city will have slightly different response sections. For example, the response section for any library near an elevator should advise people not to use the elevator in the event of a fire or an earthquake. Public libraries should include procedures to deal with bomb threats and dangerous patrons, and will probably need concise directions on how to evacuate patrons. Just as each site is unique, each response section will be more or less different from every other.

Finally, we must develop a resumption section for the library. Here we must deal with various levels of damage to the library building. If the building is unsafe to enter, where can library staff go? The resumption section should suggest alternative sites for library service—and even a bookmobile is better than nothing.

Fortunately, most disasters do not result in the destruction of an entire library building, but the resumption section must indicate the best ways to deal with severe damage. To resume operations after a serious flood, for example, it is advisable to do business with an expert service that has moisture remediation equipment and extensive experience in conservation and cleaning up water-damaged libraries. Ideally a library can reserve an expert team that will arrive at the library as soon as possible after a flood to begin cleaning up. The library should be obliged to pay for nothing until the team arrives onsite. It is important for librarians to make sure that expert teams are well equipped to handle large numbers of damp books and other materials.

Our resumption section must also address the question of lost data. This is one of the more politically sensitive issues that a library disaster planner must face. Systems librarians often insist that their data are fully backed up, and that it is simply wrongheaded to suggest that the automated catalog is a risk. But systems librarians must be made to understand that basic backup is not enough. Most libraries produce backup media, which are frequently stored in the same area—sometimes on the same shelf—as the original media. If a fire were to strike the systems department, both originals and backups could be destroyed. So what was the point in making backups?

A number of large North American libraries are currently sending their backups to secure off-site facilities for long-term storage. Various archives, record centers, book collectors, and special libraries are using the same kinds of off-site services. The resumption section will also include information regarding all insurance policies in force, and note all visits from the insurer's property inspectors. For insurance purposes, it is advisable to include a series of photographs of the library in its original condition. These photos are often the best evidence for what has been lost in a flood or fire. When the first draft of the disaster plan is complete, I advise my client to circulate it to other staff members for their comments. If there are no major oversights, we can carry out any required fire drills and orientation sessions. After that, with all due respect to my client, I usually insist that my part of the project is finished. I have done all that I can to prepare the library for those contingencies that are constantly possible, in hopes that I have helped to make their worst consequences less probable.

People, paper, data: Disaster planning for libraries

21

To lose a library is a tragedy, the extent of which we do not realize until it is too late. Consider the treasures that a major library contains. At the Library of Congress in Washington, DC, we find collections of unique books, Rembrandt prints, early American photographs and maps, antique musical instruments, and the manuscripts of various presidents. How precious is Abraham Lincoln's handwritten copy of the Emancipation Proclamation, issued in 1862 after the Battle of Antietam? How could we replace Thomas Jefferson's rough draft of the Declaration of Independence, or George Washington's 12-page ink draft of his First Inaugural Address? The loss of these items is inconceivable, but always possible.

Risks prevail at all libraries, from the largest facilities and collections in national capitals and universities, to the smallest in rural communities. Few libraries hold treasure troves as historically significant and valuable as those in the Library of Congress, but even the smallest public branch, school library, or corporate information center can contain rare materials and resources essential for research, education, and business. What disaster planning and business resumption specialists face when dealing with a library are circumstances different from those in any other institution.

Nevertheless, the development of a library disaster plan involves the same conceptual model that planners use for many other institutions. The library must be prepared for a broad range of probable and possible emergencies before, during, and after their occurrence. The safety of staff and patrons is paramount at all times; the security of buildings and collections, while vital, is secondary.

To cover contingencies before a disaster, an emergency preparedness program (EPP) is advisable. A library EPP begins with an analysis that considers regional, local, and site-specific risks. For example, the regional risks that a library faces in Los Angeles include earthquake, fire, flooding, high winds, severe smog, and civil unrest. Local risks are those prevailing in a particular neighborhood. For example, a library in a high-crime area will be more likely to sustain damage from arson and serious vandalism; a library in an outlying area might be at risk from brush fires spread by the dry Santa Ana winds. Libraries located in the coastal areas of Los Angeles must manage risks from landslides, sinking coastlines, and tsunamis.

As in most American cities, LA's general public is aware of prevailing regional and local risks. Such risks often are part of a region's history, and attract frequent media attention. Site-specific risks, however, are often unrecognized or neglected. At any library, site-specific risks arise from the age of the library building and the condition of its structural and nonstructural components. Flooding and fire are the major risks to libraries all over the world, and should be included in every library risk analysis. There are also risks from heavy furniture such as shelving, file cabinets, counters, desks, and displays. The size and weight of these objects can hinder evacuations, emergency response, and clean-up; in earthquake zones, they become even more serious hazards, along with the large panes of window glass that, under normal circumstances, provide readers with natural light.

LA librarians are wise to consider the history of disasters on their sites, with particular attention to the overall performance of their buildings during earthquakes, the age and condition of shelving units and other furniture, the availability of fire-fighting equipment, and the state of the plumbing not only in the public areas, but also in basements and staff work areas.

Human-caused risks prevail at any library that is open to the public. LA librarians are accustomed to patrons who are emotionally disturbed, drunk, or rude, but they might not be sure how to handle an arsonist or a bomber. Vandalism in the form of graffiti on a library wall does not merit disaster response, but the introduction of a virus into a library database requires immediate attention if the database is to maintain its integrity and accessibility. LA librarians will note that human-caused risks prevail at some sites more than others, and should insist on effective planning and procedures that address the risks at each specific site.

The second part of a library's EPP is the mitigation program, the goals of which are risk reduction and loss prevention. Ideally this program will either prevent a disaster from occurring, or—in the case of a large natural disaster—lessen its effects when it strikes.

Mitigation at any site entails training. Every librarian should take a first aid course and know how to use a fire extinguisher. Every librarian should know how to manage an evacuation, handle a bomb threat, and call for the appropriate first responders. In some libraries, comprehensive orientation and training programs are open to all employees; other libraries depend on the personal initiative of employees to prepare themselves and the sites at which they work. The most successful mitigation programs include well-organized training for all staff. Library directors and other senior managers are encouraged to show their support for the program by undergoing the training themselves.

Preventative maintenance is another essential element of a library's mitigation program. Floods and leaks can be prevented through regular inspections of the plumbing, roofs, and windows. Frayed and faulty wiring should be replaced as soon as possible in order to mitigate the risk of fire. Since flooding and fire are the most common disasters in libraries, a preventative maintenance program that addresses these risks effectively will reduce the worst losses in most libraries.

Libraries in earthquake prone zones must consider the risk of falling shelves. The best way to prevent a shelf from collapsing during an earthquake is to moor it to the floor. Unfortunately, many library floors are not built to accept the addition of mooring, which includes large bolts and anchor panels. In older library buildings, shelves are often supported by cross-bracing and metal cables or strips running from the top of one shelf to the next. While shelf-top supports will hold up shelving during a small earthquake, they often cannot sustain the pressure generated by a moderate or large earthquake. To mitigate this risk, libraries might have to consider a shelving replacement campaign. This is an expensive solution, but it is easier to accept before the shelving has collapsed than afterwards. Moreover, on some sites the replacement of rickety shelves is the only way to reduce risks to library employees and patrons. When a shelf weighing several tons collapses on a hapless researcher, the consequences can be fatal.

Another problem in earthquake zones is shattered glass. Modern libraries often contain huge windows that brighten the decor and create a pleasant work environment. Glass walls provide a sense of openness as well as sound control and privacy. To mitigate the risk from glass that shatters during an earthquake, librarians should consider barriers such as venetian blinds or curtains. Even a thin gauze curtain can stop a large quantity of glass debris from flying into a public area or busy work space, and does not have to detract from the atmosphere in normal circumstances.

Mitigation measures for library computer systems include regular backup of essential data, as well as up-to-date firewalls for all library databases accessible through the Internet. Most card catalogs have disappeared. Now libraries provide online systems that are more convenient and less labor-intensive. But while an electronic catalog is a wonderful research tool, it is vulnerable to a variety of risks that demand a carefully organized backup routine and the librarians' vigilance.

Hackers are a growing risk to library systems, but librarians should remember that much damage to systems is due to simple accidents. Planners refer to the "Whoops!" factor when they describe an incident in which a librarian trips over a cable and disables an entire library network, or knocks a PC off a counter and loses 3 years of work, or accidentally activates the sprinkler system in the computer room and drenches a new set of servers. Such accidents are common; fortunately, serious losses can be avoided through simple precautions. Cables should be laid in such a way that people cannot trip on them; all hardware should be moored to supporting surfaces; and any data stored in hardware located near a sprinkler system should be backed up at least daily. The best way to guard against the "Whoops!" factor is to apply common sense to human weakness at every opportunity.

Common sense also demands appropriate emergency supplies at each library site. Librarians will agree that a first aid kit is essential at any busy public site, but a kit is useful only if its user is skilled in the application of first aid. A pamphlet concerning emergency measures for heart attack victims can be informative, but library employees should not need to review it while a patron expires beside the reference desk. As the Baby Boomers grow older and become more susceptible to ill health, those in charge of public facilities such as libraries will be increasingly obliged to turn the first aid kit into a life-saving tool by taking the necessary first aid courses. The first aid kit cannot save lives by itself.

Other emergency supplies include fire extinguishers, flashlights, and battery-operated radios. Library employees should never have to spend time finding these items: these should always be close at hand. A librarian at each site should make sure that the fire extinguishers are fully functional and regularly inspected. New batteries should be available at each site, and stored in an obvious place rather than hidden.

Crow bars and shovels, dust masks, blankets, distilled water, dried foodstuffs, hard hats, and body bags have also been stored in libraries for emergency purposes. In the end, the most useful supplies are those that are indicated by a library's risk analysis. Librarians should ask themselves what tools they will need directly after any emergency indicated in the analysis, and provide their sites with the appropriate supplies.

Once the EPP is complete, librarians should study it carefully and revise it regularly. EPPs for public and academic libraries should be distributed to civic authorities, library trustees, and senior administrators outside the library system whose responsibilities affect library operations. Corporate (or special) library plans should be distributed to all senior management outside the library and to any custodians or property managers who are involved in the library's preventative maintenance.

Protecting lives during a disaster is the main purpose of the library's disaster response program (DRP). A practical DRP for any library includes one brief document and a series of drills. The document gives site-specific, point-form directions on what library employees should do in the event of any of the disasters mentioned in the risk analysis. These directions are clear, concise, and life-saving. Usually they do not cover any but the most crucial instructions for human safety, although some libraries use the DRP to promote first aid courses and specific safety measures, and to remind all library employees to review the EPP and DRP regularly.

The DRP document can be a single page suitable for posting on bulletin boards, or it can be folded into a small-format brochure for distribution to every library employee. It is advisable to make the DRP document as attractive as possible, so that employees understand its importance. An informal, hastily produced DRP will not have the same impact on employees as one that is properly laid out with the library's logo and site address(es).

Once the DRP document has reached all employees, drills are necessary to reinforce its directions. The library director or a senior library manager should ensure that earthquake and evacuation drills are regular and efficiently executed. It must be emphasized to all participants that evacuation drills are not excuses for a cup of coffee down the street or a late-morning smoke in the parking lot. Once the alarm has sounded or the order to evacuate has been made, employees should proceed to the previously arranged safe gathering site without delay. Patrons in the library at the time of the drill should be asked to leave the building temporarily, with apologies for any inconvenience. When all staff members have assembled at the safe gathering site, the time required for the full evacuation should be noted. A de-briefing session should follow soon afterwards, during which evacuation participants should mention any impediments along the evacuation route or other concerns about the effectiveness of the drill.

The final phase of disaster planning involves the formulation and testing of measures to restore facilities and to reopen the library for service. A comprehensive business recovery program (BRP) gives library employees at all levels practical directions to deal with postdisaster circumstances.

A BRP begins with advice on matters that seem governed by common sense: how to recognize and declare a disaster. In a library, however, disaster recognition can be more complex than at other institutions. For example, if a large public or academic library loses its online catalog, it is out of business, at least for purposes of reference and research. Loss of the catalog is one of the more serious emergencies that can strike a library. There is little point, however, in calling 911 when a systems crash causes data loss. Some disaster planners would not consider such an event a true emergency. But librarians, realizing that their ability to offer service depends on the accessibility of the catalog, regard data loss with alarm.

The BRP acknowledges all risks that librarians regard as threats to their operations, whether those risks be general (fire, flood, etc.) or library-specific (catalog data loss).

Some library BRPs contain analyses of contingencies with library-based examples. For example, the least serious event that a BRP might deal with can be described as a "serious incident," which could involve a minor loss of data, a roof leak that drenches several shelves of replaceable books, or a threat from a drunken patron. An "emergency" could be used in the event of a single casualty, a moderate fire, or substantial vandalism that compromises security at a site. A "major emergency" covers serious damage at a single site, and possibly several casualties.

A "disaster" is defined as an event that is beyond the powers of first responders to prevent or control, and that results in serious damage and prolonged service disruption at several sites and possibly a number of casualties. Examples are large floods and moderate earthquakes. Disasters that cause horrendous, region-wide property damage and loss of life are identified as "catastrophes." Fortunately, catastrophes are uncommon, although they remain indefinitely in the public memory. Hurricane Katrina, the Kobe earthquake, and the Chernobyl meltdown are outstanding examples.

Once librarians can identify with precision the magnitude of an event, they can make an informed decision as to the necessity of making a declaration. Sometimes the disaster declaration involves little more than the director's announcement to staff, a call to the chair of the library trustees, and perhaps a press release. In other instances such as a system crash, an announcement to staff might be all that is necessary or appropriate. As soon as the director makes some kind of declaration, however, the library's emergency management program (EMP) commences.

Library EMPs cover safety of employees and patrons onsite after an event, damage assessment procedures, the roles of the security officer and building maintenance managers, the securing of damaged sites, and safe travel measures to and from damaged sites. EMPs also contain sections on postdisaster communications such as press releases, emergency signage, and the use of couriers. The purpose of the EMP is to stabilize a difficult situation and to assist the library staff in retaking control of operations at any site affected by an event. Once calm prevails and the library is secure, the librarians can begin basic clean-up procedures.

As always, however, the library's computer department requires special attention. The BRP should describe the library system's minimal configuration: the list of hardware and software that must be available and functional for the computer department to offer its basic services.

Damage assessment in a computer is a specialized concern, and is often left to the computer vendor who holds the service contract. If the computer room has been destroyed, the BRP outlines the necessary steps for the activation of the library's remote site. While smaller libraries find remote sites beyond their budgets, they can often make arrangements with another institution for anything from backup tape storage to the temporary use of the institution's systems.

In an age of shrinking budgets and rising costs, libraries are exploring the possibility (and, in some cases, reaping the benefits) of partnerships with external agencies. In fact, the development of its BRP gives the library an opportunity to form strategic partnerships with the broadest variety of agents: computer vendors, clean-up and restoration firms, building contractors, engineers, publishers, booksellers, and conservation specialists. The idea behind every partnership is that the library will have its partners ready and able to provide services and replacements for anything from shelving to paperback fiction. Should a priceless manuscript be soaked in a flood, the BRP will contain the address and telephone number of the appropriate conservator who can repair the damage. Sometimes libraries will include written agreements from agents, who state that, in the event of a disaster, the library will receive the agents' attention before any other institution.

Even libraries that employ full-time and highly specialized conservators will form partnerships with outside conservators, in case a disaster overburdens library staff. Since library clean-up is extremely labor-intensive, it is wise to have reliable support available.

Other BRP sections deal with insurance, restoration of facilities, and issues arising from disrupted work schedules, overtime, and reallocation of staff to library sites that need more assistance with clean-up. Getting a library back to normal can be costly and time-consuming, but with a well-organized BRP, recovery will not be such an arduous process.

Sadly, library history contains accounts of hundreds of disasters that have wiped out almost entire fields of knowledge. From the destruction of the Library of Alexandria in the third and fourth centuries AD, to the incendiary bombing of the Bosnian National Library in 1992, the inventory of losses might make us think that the effort to save libraries and their contents is useless. But that effort, conducted with increasing sophistication by disaster planners and recovery specialists in the library community, has saved enormous amounts of treasure from oblivion. More optimistic versions of history will record not what has been lost, but what has been rescued from fire, floods, earthquakes, war, and human carelessness. With the planning methods that are available to us, the future for libraries is brighter than we might think.

21.1 A sample library DRP

This sample DRP is similar to several in effect for libraries on the West Coast, from British Columbia to southern California. Note that earthquake instructions are not as necessary in Florida or the Carolinas as instructions regarding hurricanes. Every DRP should address its appropriate regional risk profile, which is that series of risks that are most probable in a particular region. Local and site-specific risks should also be covered in a DRP.

(Library logo)

Paradise County public library

The librarian's personal DRP

- Fire
- Flooding and leaks
- Explosion
- Bomb threat
- Earthquake
- Toxic spill
- High winds/severe weather

21.2 Fire

- Small fires can be doused with fire extinguishers. If the blaze starts to spread, trigger the fire alarm.
- Main library: Advise the security officer. Prepare to evacuate on command.
- Branches: Evacuate immediately.
- Walk, don't run.
- Avoid flames, fumes, and smoke. The closer you are to the floor, the lighter the smoke and fumes.
- Do not use the elevator.
- Advise patrons to leave. Do not remain behind if a patron refuses to leave.
- Once outside, go to your safe gathering site: (location).
- Upon arrival at your safe gathering site, the senior employee present will make sure that the fire department has been contacted.
- Branches: Once you are sure that the fire department has been contacted, call the security officer at the main library.

21.3 Flooding and leaks

Most floods and leaks are not life-threatening, but observe the following:

- Avoid wet wiring and electrical outlets. Advise patrons to stay clear of computer terminals.
- Do not walk through water.
- Try to identify the source of the water, roof, window, sprinkler, pipe, washroom, etc.
- If flooding is caused by faulty plumbing, try to turn off the water main. Contact building maintenance without delay.
- Do not begin clean-up until after an inspection of wiring and electrical outlets.

21.4 Explosion

- Main library: Advise the security officer immediately. Prepare to evacuate on command.
- Branches: Evacuate immediately. Upon arrival at your safe gathering site, call the fire department and the security officer.

21.5 Bomb threat

By telephone:

- Cooperate with the caller. Listen carefully.
- If time permits, ask the caller:
  1. the location of the bomb
  2. the expected time of the explosion
  3. why the bomb has been planted.
- Try to remember the exact time of the call.

- Try to remember the caller's voice: male/female, accent intoxicated, familiar?
- Note any background noises during the call: traffic, construction, music, voices, etc.
- Immediately advise the senior staff member at your site.
- Main library: Senior staff member in the affected department should call the security officer immediately and prepare to evacuate on command.
- Branches: Senior staff member should briefly check work areas for unusual or out-of-place items. Do not touch any suspicious parcel, envelope, box, or container. Then evacuate the branch. Take all personal belongings with you to the safe gathering site.
- As soon as evacuation is complete, call 911 and the security officer at the main library.

In writing:

- Handle the document as little as possible after you have read it.
- If it is attached to a wall or other surface, do not touch it or the surrounding area.
- Advise the senior staff member at once.
- Main library: Senior staff member of the affected department should call the security officer to notify the police. Prepare to evacuate on command.
- Branches: Senior staff member should call 911 and evacuate the branch immediately. Upon arrival at your safe gathering site, call the security officer at the main library.

21.6 Earthquake

During the shaking:

- Protect your head. Take cover under a counter, table, or desk.
- Avoid shelving units, file cabinets, glass, loose masonry, and utility wires.
- If indoors, stay indoors. If outside, stay outside.
- If in a moving vehicle, stop in a clear area away from falling debris. Do not leave the vehicle.

After the shaking:

- Apply first aid if necessary.
- Avoid elevators.
- Do not use candles, matches, or other open flames. Do not smoke.
- Do not make telephone calls unless they are life-saving.
- Turn on your radio or television for emergency bulletins or updates.
- Avoid entering damaged buildings.
- Expect power outages.
- Expect aftershocks.
- Sprinklers and alarms might be triggered.
- Be prepared to remain at your library site for up to 3 days.

After 6 hours:

- Call your pre-arranged, personal out-of-state contact number. Give your name, local time of your call, location, and status.
- Branches: Senior staff member should inspect the branch for damage. If you suspect that it is unsafe, evacuate to your safe gathering site. Do not attempt to move fallen shelving without assistance.

21.7 Toxic spill

Most toxic spills near local library sites would occur on streets, particularly major arteries.

- Avoid the spill as much as possible.
- Main library: If you notice heavy fumes, call the security officer and prepare to evacuate.
- Branches: If you notice fumes or any toxic material seeping into the library, evacuate to the safe gathering site. Call 911 and the security officer at the main library.
- Do not use matches, candles, or other open flames. Do not smoke.

21.8 High winds/severe weather

- Expect power outages and telephone line disruptions.
- Avoid glass.
- Main library: Prepare to close the affected departments upon the instructions of the director or security officer.
- Branches: Wait for instructions from the director or security officer. If contact with the main library is lost, determine if it is safe to evacuate the branch.
- If in a vehicle, drive carefully. Slow down. Avoid fallen power lines and wires, trees, and unsafe structures.

21.9 General precautions

- Take first aid training. Keep your ticket up to date.
- Make sure that you know the locations of the nearest first aid kit, flashlight, and fire extinguisher.
- Familiarize yourself with the nearest safe exit(s) and safe gathering site.
- Participate in all drills.
- Keep a spare pair of shoes and eyeglasses at your workplace.
- Discuss safety measures with all members of your household. Make sure that everyone knows the out-of-state contact number.
- Review all safety procedures every 3 months.

If you have any questions or comments, contact the security officer. Extra copies of this DRP are available from the director's office.

Let's stay safe in paradise!

Investigating risk: Assessing and analyzing trouble before it strikes

22

22.1 Risky business

Your director asked me to visit your branch and have a look around. While I'd like to browse in the stacks and find myself a couple of decent mysteries, I'm here to perform a risk assessment and analysis. I'm not a police officer, and nobody's in trouble—at least not legal trouble. I'm a risk manager, and I want to investigate the risks that prevail on your site. With luck and a clear head, I'll identify anything that might cause you grief.

If you'll excuse me, the first thing I must do is to walk out through your front door and around the site perimeter. I'm looking for proximity risks, anything in the vicinity that might affect your operations. For example, the self-serve gas station across the street could catch fire or explode, or its tanks could leak. Leaves could plug the drains at the side of your building and cause a flood. The bakery on the corner could catch fire. Finally, I don't like the look of that abandoned office block down the street. It's an old building, and could attract an arsonist.

22.2 Smoking is bad for you

Do I sound like Chicken Little? Perhaps. But it's my job to spot risks before they can hurt you and your staff, and disrupt your operations. If I can pick out the potential trouble spots, then you can do your job without having to worry about nasty surprises. Fact is, many disasters arise from a chain of tiny events. You see that young man at the gas station? He has just filled up his 4×4 with premium blend. Before he drives off, he lights his cigarette and tosses the match out his window. It lands in a pool of gas that somebody spilled a few minutes ago. We have an ignition! While the Fire Department will arrive quickly and douse the flames, and nothing else explodes, there is smoke damage in your branch. The rugs and curtains need cleaning. Maintenance crews should check the ventilation system and change the filters. While they're on site, you must ask them to clean the windows, which are coated with smoke residue.

Unfortunately your computer hardware has been contaminated with smoke particles, and there is a high risk of system failure. You have no choice but to shut down every terminal and wait for a specialized crew to clean the affected components. Your patrons are not pleased, and you'll spend a lot of time explaining to them why they can't use the OPAC, why various areas must be closed for cleaning, and why everything smells smoky. Unfortunately, the odor of smoke can persist for weeks or months after a fire, even after cleaning is complete.

All of this damage occurs within a few minutes. Much of it is preventable. As long as you recognize what might happen, you can mitigate any risk that prevails on your site.

22.3 Assessing risk

We can't analyze the risks at your branch until we determine what they are. Usually we divide them into four main kinds: natural, human-caused, technological, and security-based. The imaginary gas station explosion is an example of a human-caused risk. People may be model citizens with the purest intentions, but they can make mistakes that lead to serious damage and fatalities. Natural risks include high winds and winter storms, earthquakes, floods, and forest fires. Librarians in Central Canada know all about winter storms, especially those that feature freezing rain and power outages. In British Columbia and parts of Quebec, earthquakes are a constant threat. The librarians of Winnipeg worry about floods, particularly those that could inundate their downtown neighborhoods. Technological risks arise from complex systems that break down: the fate of any system if it functions for long enough. Think of systems crashes "of obscure origin," certain kinds of power failures, downed airliners, and air conditioning systems that malfunction during heatwaves and turn your Technical Services Department into a sauna.

Security-based risks include everything from theft and vandalism to sabotage and bomb threats. These are human-caused risks that prevail owing to criminality rather than clumsiness. Many libraries focus on security at the expensive of other kinds of risk management. Your branch may employ Tattle-Tape, alarms and other security measures to reduce theft of valuable materials, but do you have ways to treat moisture-damaged books? (By the way, that skylight above your Fiction section looks old. Has anyone inspected it for cracks?) And what would you do if a fire broke out in the basement?

When assessing risks, you should identify all of those that prevail on your sites. This does not oblige you to take into account every possible problem; it is reasonable to disregard unlikely threats such as meteorites and circus animals on the loose. If your branch is in Regina, you needn't worry about tidal waves or avalanches; if you work in Toronto, your risk assessment needn't include megathrust earthquakes. Common sense and a knowledge of your site's history will give you much of the information you need for your assessment. In many libraries, the greatest risks have already made themselves known.

22.4 From assessment to analysis

Many risks lead to water damage, which is the most common result of natural disasters in North American libraries. Heavy rains in Vancouver penetrate an old roof and drench a library's magazines and reference area before seeping into the basement and soaking materials in storage. Melting snow on the Prairies will cause similar problems, along with slip-and-fall injuries in the parking lot. High winds in Ontario cause cracks in roofs and building exteriors that allow water to leak inside. In Quebec, a river overflows its banks, submerges a town site, and forces librarians to watch their books act like sponges. A combination of wind and water is always hazardous to libraries in the Maritimes. No matter where your library is located, it is wise to mention "water ingress" as the first and most likely risk. Ironically, the effects of a roaring blaze are often not as bad as the torrent of water from the sprinklers or firefighters' hoses that extinguish it.

Human-caused risks in libraries are closely related to technological risks. In essence, when people make mistakes in libraries, they cause damage or loss mostly through the misuse of technology. Accidental erasure threatens every database. Operator error leads to the introduction of viruses into the software containing borrowers' records. Or somebody forgets to check the computerized lock on the main entrance, and one night the library is left open. Opportunistic thieves discover how easy it is to steal terminals and CDs. Perhaps they trigger the alarm system, but there are a number of cases in which contract security guards ignore the signal. With respect, one of the biggest risk factors in any organization is the staff. But that's between you and me—and anyone else responsible for assessing risks.

Security-based risks include threats to the safety of staff and losses from vandalism and theft. Some sites are more prone to hostile patrons and intruders than others. Most reports of workplace violence in libraries come from sites in large cities, although rural branches are not immune to bomb threats and threatening behavior at the reference desk. City libraries also report more theft and vandalism. While it is impossible to eradicate security-based risks, it is foolish to ignore them, or simply to accept security breaches as they occur. To mitigate any risk, the first step is to identify it, then determine when and where it is most likely to happen, and what the most likely results will be. Hence the listing of risks in an assessment conduces to an analysis of those risks, and might suggest ways to reduce them.

22.5 What's the likelihood?

The risk assessment for your branch lists water damage at the top, followed by fire and smoke, power outages, and winter storms. Based on the branch's history, your main security-based risks are theft and vandalism. Your next step is a risk analysis, which will determine how likely it is that these risks will disrupt your operations. The main source of unwanted water will be that skylight, and with our region's increased frequency of rainstorms, there is a strong likelihood that someday the old insulation will give way. A leak is the result: that's the commonsensical analysis of this particular risk. To mitigate it, you should check the insulation and replace it if it can no longer keep out the rain.

Fire and smoke can be life-threatening, and thus demand more vigilance. May I suggest that you consult your local Fire Department for advice? They will check your fire control equipment and make sure that you comply with your local Fire Code. They will also tell you to remove the pile of discards from the stairwell, and to hold the evacuation drill that you have postponed for the past 5 years. Power outages and brownouts are likelier during periods of inclement weather. To prepare your branch, consider an uninterruptable power supply such as an emergency generator, and make sure that you have flashlights available at key points such as the reference and circulation desks. Winter storms can cut your power and cause build-ups of snow and ice around entrances. With climate change, these storms are not as frequent as they used to be, but they're still a threat.

22.6 Dealing with villains

Your security measures are sound, but your theft rate continues to rise. I suspect that thieves are removing Tattle-Tape in the washrooms, and walking past the sensors without triggering the alarms. You can mitigate this risk by posting signs in the washrooms: *For your convenience and safety, security guards patrol all areas of the library.* The message is clear and threatening to nobody but the thief in the cubicle who's trying to remove Tattle-Tape with a pair of tweezers… Is that the security guard who just walked in? Is he looking for me?

Library vandalism is an expensive nuisance that occurs in the form of book mutilation and graffiti on washroom and exterior walls. To mitigate this risk, librarians should post notices announcing frequent patrols, and asking the public to be on the lookout for graffiti artists. To ward off book mutilators, you can let patrons know that damage to items will be investigated and traced, and that those who damage books can lose their borrowing privileges. You can't wipe out security problems, but you can scare off or discourage potential perpetrators.

22.7 Be site-specific

Every site is unique, and each risk assessment and analysis comprises different sets of risks at varying levels of likelihood. Moreover, risk profiles change over time, as buildings age or undergo renovation. That's why risk management is an ongoing process, for which you and your colleagues must take responsibility. With so many sites to inspect, I can't drop in regularly. If you order the latest titles by Ruth Rendell and P.D. James, however, you might see me more often than most.

A van and a plan: How consortium offices can contribute to disaster recovery

23

23.1 Scenario: Stormy weather

You are the head of a suburban public library branch. At 5:00 a.m. a storm strikes your community. High winds cut power lines, and the branch will remain dark and cold for at least 2 days. Heavy rainfall seeps through your leaky roof, resulting in pools of water around the reference desk and in the Children's section. More than 800 hardcovers are soaked, along with hundreds of softcovers and paperbacks.

At 8:30 a.m. you arrive at the branch. You attempt to telephone the Library Administration Office to discuss matters with the director, but the line is dead and will be out of order for the next 12 hours. (The director will not reach her office for another 4h because a tree has fallen on her garage, collapsing its roof onto her car. Local bus service is irregular. In the end, she walks to work.)

You turn on the radio—it is battery-operated, isn't it?—in the staff room. You listen to reports of flooding, road closures, and casualties. At 9:05 a.m. a woman and her little boy arrive at the front entrance. The woman tells you that her house is flooded and she has nowhere to go. The boy states his preference for stories about elephants.

23.2 What now?

Few libraries maintain effective disaster plans. When a regional disaster occurs, they can encounter a variety of serious problems. Unfortunately they lack the resources necessary to resume operations quickly and safely. Library personnel are not often trained to deal with or work around damaged facilities. Moreover, libraries are a low postdisaster priority in most municipalities. Not surprisingly, schools, hospitals, and seniors' centers merit the attention of recovery crews before any library facility.

Frequently after a disaster, librarians find themselves out in the cold—literally. They complain that it is difficult to return to work when so little emergency support is available. In particular, they face problems with communications, transportation, safety inspections, IT system restoration, and public relations. To wait for municipal or regional governments to offer help can be frustrating. What governments eventually provide can be too little, too late.

Robertson on Library Security and Disaster Planning

23.3 Desperately seeking assistance

In the past, librarians have tried to alert trustees and government agencies to the risks that prevail in their communities and the potential effects of a disaster. Board minutes contain numerous mentions of the need for emergency preparedness and response procedures, business resumption plans, and postdisaster safety measures at library sites. But these minutes do not necessarily lead to action. Boards assume that governments are the only source of assistance, and that libraries must wait at the end of a long queue.

Recognizing their precarious situation, some librarians have attempted to draft disaster plans. Generally these documents concentrate on preserving collections and restoring IT systems. Occasionally they deal with other problems such as emergency moisture control and damage to window glass and shelving. But many plans do not refer to external agencies that could offer valuable help. An example of such an agency is your local library consortium office.

23.4 Doing what comes naturally

As organizations, consortium offices are usually much smaller than any of the libraries they serve. Most operate from a single site with only a few employees. Nevertheless, consortium offices can have a perspective on local public library operations that is broader and more detailed than that of any single library in the area. Just as librarians know their patrons' interests and tastes, consortium administrators are aware of what their member libraries need and want. Consortium office staff can be as politically connected as any group of library managers or trustees. They can also be good communicators who consider the timing and content of every announcement, message, and memo that they send out to their library communities.

These attributes are especially useful after a regional disaster. At present, however, consortium offices in most jurisdictions rarely become involved in library disaster response and business resumption processes. Consortium administrators have no wish to be compared to the volunteers who loiter in a disaster zone hoping to be given something useful to do. As disaster relief workers know, such volunteers can be more of a burden than a help. Hence consortium offices usually wait for libraries to reestablish communications and standard working relationships. And libraries do not often request extra or different services from consortium offices after a disaster.

23.5 Consortium offices first

Consortium offices should develop their own disaster plans before becoming involved in a library's postdisaster activities. Consortium administrators will be relieved to know that a consortium office's disaster plan is usually easier to compile than a library system's. Consortium offices have less of everything, including staff, physical assets, and data. Consortium employees do not occupy high-traffic space or deal directly with thousands of patrons every day. Consortium offices can be shut down, relocated, repaired, and re-opened much more quickly than a library. Small turns out to be flexible and adaptable as well as beautiful.

Disaster plans for consortium offices should include human safety procedures appropriate to the site and its prevailing risks. Simply stated, consortium employees should know when and how to leave their offices safely. They should back up vital data for purposes of in-house business resumption and service to libraries. Emergency communications and transportation procedures should be in place. First aid equipment and expertise should be available for all employees. And if a consortium office is destroyed, its staff should have an alternative site where they can resume operations.

23.6 Postdisaster services to libraries

Consortium offices can offer a variety of crucial postdisaster services to their member libraries. Since emergency communications and transportation are essential to any organization's business resumption and recovery processes, a consortium's humble van can play an important role in helping libraries to move anything from books and IT equipment to furniture and even employees. Postdisaster transportation can involve moving assets from a damaged facility to another site, such as the closest branch or storage facility.

Consortium offices that manage interlibrary loans can also reroute deliveries to alternative sites and act as couriers. After regional disasters, telephone systems might be out of order for days. Commercial couriers are overloaded. The only alternative for libraries might be the consortium van, which can carry messages and other items between sites as long as it has a supply of fuel.

Larger library systems may maintain fleets or trucks and other vehicles, but the consortium van can render invaluable services across an entire region. Often the consortium driver is familiar with more routes across the region than a driver who covers a single municipality. In difficult and sometimes dangerous circumstances, the consortium driver's experience and road knowledge could be an important contributor to a library's business resumption.

23.7 Additional services

Depending on their relations with local governments, consortium offices could also facilitate other postdisaster tasks such as safety inspections and IT system restorations. Municipal engineers can be inundated with requests for inspections of public buildings and sites. Libraries will probably be low on the inspection list. A consortium manager, however, could keep in touch with municipal engineering departments and offer useful information regarding local library buildings that appear to have sustained significant damage. Engineers appreciate such information, since it can save time. They don't need to communicate with each local library system and visit every library building.

Another service involves working with IT vendors to replace computer equipment that has been destroyed. Hardware is notoriously fragile. After disasters, insurance adjusters will often write off components rather than performing lengthy tests of their functionality. Consortium offices could act as clearinghouses, particularly when out-of-town vendors have difficulty delivering replacement equipment to sites that have been shut down after a disaster. If a vendor cannot drop off those new workstations at a central library that is "closed until further notice," a consortium office could store them temporarily.

23.8 Public relations in hard times

When library directors are busy allocating staff to different branches and trying desperately to find contractors to repair moisture-damaged buildings, consortium offices could assume responsibility for public relations tasks. These include issuing media releases and public service announcements regarding temporary closures of libraries and alternative drop-off sites for library returns. Through media interviews, consortium spokespersons could reassure the public that library service in the region will be restored as soon as possible, and that library staff are looking forward to seeing their patrons again in the near future. Those patrons who consider the library to be an important part of their personal routines will welcome such reassurances.

Consortium offices could also deal with volunteers and their allocation to damaged sites. While many libraries prefer to control volunteer efforts from their main administration centers, sometimes there are simply too many volunteers to manage. They could even endanger themselves by trying to enter crumbling or smoldering buildings. Library administrators who have handled postdisaster volunteer "convergence" admit that they would have welcomed assistance from other libraries or affiliated organizations such as consortium offices.

23.9 In a pinch ...

Ideally there would be no need for a consortium office to assist a library in its disaster recovery efforts. A perfect world would not include high winds, fires, floods earthquakes, ice storms, and power outages. Experience tells us, however, that such events are always possible, and that libraries are more vulnerable than many other public institutions. It would be prudent, therefore, to consider ways in which consortium offices could contribute to their respective regions' disaster recovery. Those offices might need plans to protect themselves and policies to describe the postdisaster services that they could provide. Of course we hope that such plans and policies need never be activated. But if a storm forces the closure of library facilities, consortium offices should be prepared to offer assistance.

And when the wind dies down, we can return to stories about elephants.

Lights out! Dealing with power outages in your library

24

Archeologists claim that prehistoric evenings included tribal gatherings around bonfires. A crackling blaze provided ancient peoples with warmth, cooked food, and opportunities for gossip and storytime.

Occasionally, however, firewood was unavailable. The resulting lack of fire led to frigid huddles in pitch-black caves, and attacks by wild beasts emboldened by the darkness. For long stretches, life consisted of little more than a dreary wait for daylight or a series of defensive preparations in anticipation of nightfall.

Fortunately we live in more predictable times—or so we like to think. Generally we're safe in assuming that the sources of our heat and light will continue to operate without interruption. But when a power outage occurs in our libraries, we're surprised and often unready for the immediate effects.

Outages can be due to different causes. High winds and winter storms damage power lines and poles. Floods and forest fires disrupt power transmission across regions and lead to prolonged outages. Sabotage of power supplies is another possibility. Every week in some North American city, a driver loses control of his or her vehicle, rams a power pole, and sinks an entire neighborhood into temporary darkness. Like every other building in that neighborhood, the local library blacks out.

24.1 Darkness visible

What happens inside a library during an outage? Suddenly heating and air conditioning turn off. Lighting and computer screens flicker, then die. The librarian opens the drawer where the flashlight should be, only to discover that it's not there. A patron stumbles toward the reference desk to complain that without a working desk lamp he can't read the small print in an encyclopedia. A woman asks the circulation clerk if she'll have to pay fines on books that can't be discharged because of the outage. Meanwhile her young sons pretend that the children's department is a haunted forest, and go howling down the aisles.

In the staff room, the supervisor finds the flashlight in the cutlery drawer. She descends the stairs to the basement in search of the fuse box, which resides behind a mountain of boxes crammed with discards. Before she uncovers the fuse box, the librarian will tell her to lock up and go home. There's no point in trying to work in a dark space thats quickly growing colder. Or hotter and stuffier, if the outage happens in summer. Besides, fuses aren't the problem.

Panic during outages is rare. In fact, most patrons will agree to leave the library as soon as they realize that power will not be restored for an extended period. Since there's a risk of slip-and-fall injuries as people move from stacks and public areas toward exits, they should proceed carefully. Library employees can act like theater ushers, using flashlights to point out safe exit routes.

Once patrons have left the library, employees can post the necessary closure signage and lock up the building. The librarian should contact the director and let her know that the library has closed early. Radio bulletins will announce how long the outage might be.

Such is one scenario. There are others. For example, if the outage occurs during daylight, and the indoor atmosphere remains stable and comfortable, the librarian may decide not to close. Illumination from windows might allow patrons and employees to continue their activities. Safety is the primary concern in this instance. If the available light is strong enough to help prevent injuries from falls and collisions with furniture, the library can stay open.

24.2 Backup generators

Many large libraries depend on generators for back-up power. Ideally a generator will start up as soon as its sensors detect an outage. To do so, it must undergo regular maintenance and testing, and its fuel tank should be kept full. Under these conditions, the library can rely on back-up power for several hours or more, depending on the size and capability of the generator.

Unfortunately, some libraries neglect their generators. Fuel runs low and tanks are almost empty when they're needed most. Servicing is infrequent and cursory, and various components can fail just as the generator starts to operate during real emergencies. There's no point in installing a generator unless library management is willing to fund its maintenance, and custodians are prepared to carry out regular servicing.

Reliable generators are especially necessary in libraries that operate elevators, escalators, and conveyor systems. The failure of these devices can lead to more than inconvenience. Patrons trapped in a dark elevator for an hour or more will not be inclined to view the experience favorably. Employees forced to stand around and wait for a conveyor to start working might start doubting the advantages of technology, especially when faced with substantial backlogs and overtime.

24.3 Electric doors

Another scenario involves the library during nonbusiness hours. If the outage occurs at 3:00 a.m. on a Sunday morning, what might happen? In many libraries, nothing. In some, however, there could be security problems, especially if alarms and electric door systems lose power. Some electric doors will jam shut in an outage. It's necessary to override the electrical system that controls their movement in order to pass through them. During outages, some library employees have simply pried open electric doors. So have burglars.

24.4 Let there be light

Librarians need little to deal effectively with outages. First, they must accept the possibility that there could be at least one outage in their building's future. In North America, the risk of an outage is higher than that of a fire, bomb threat, or earthquake, and considering the recent warnings of civil authorities about the potential for regional grid failures, librarians would be wise to prepare for them. Ignoring the risk is not an adequate response.

Every library should have a supply of flashlights: not just one, but several. These should be stored in conspicuous locations. In staff rooms, cafeterias, and offices, flashlights can be mounted on walls. They can be stored in the top drawer of the reference desk, in the first aid kit, and in the staff washroom. A good supply of spare batteries should be handy. Every 3 months, a library assistant should be assigned the task of making sure that all flashlights work.

In a library with a back-up generator, a librarian should be responsible for ensuring that it is serviced regularly, and that its fuel tank remains topped up. The librarian should keep a record of all servicing, refueling, and testing, and work with custodial staff to establish these activities as standard operating procedures.

Circulation counters should contain materials for making temporary signage to be posted on front entrances. White cardboard, duct tape, and a black felt pen are all that library employees need to make an effective emergency closure notice: *Closed due to power outage. Sorry for any inconvenience.*

If an outage forces you to evacuate your library, you should have procedures for clearing public and staff areas with nothing more than a flashlight and a loud voice. It's helpful to test these procedures. How? One evening after you've closed your library, ask a couple of colleagues to stay behind for a minute or two. Then turn off all the lights. You'll quickly understand how an outage affects your workspace. Have you ever moved around the staff room, office, or stacks in the dark? It can be an unnerving experience, but you will automatically appreciate the benefits of a flashlight, as well as the potential dangers of common objects such as stepping stools, waste-paper baskets, and area rugs, which seem to have been designed for you to trip over in the dark.

24.5 On the bright side

If predictions of climate change are correct, and we can expect frequent storms and grid failures, then we should prepare for what some emergency planners refer to as the New Dark Ages. Happily most outages will be brief, and lead to nothing worse than minor inconveniences such as spoiled food in staff room fridges and temporary closures. Some outages might have longer durations and result in IT failures, data loss, and injuries. Nevertheless, a few simple preparations can prepare you and your library for any sort of outage.

Count yourself lucky. The only wild beasts you might encounter are the young patrons in that haunted forest.

Water finds a way: Dealing with leaks and floods in your library 25

Ask Noah. Most people don't take water damage seriously until it's too late. Libraries are especially vulnerable. Every morning, across North America, librarians arrive at work to discover pools of water in their basements, soggy carpets in their reference areas, and tons of damp periodicals and books.

Why are library leaks and floods so common? Partly because there are so many sources of water. Plumbing cracks. Sprinklers activate by accident. Roofs wear out, especially when we include rooftop parking in our building designs, and the weight of countless vehicles causes fissures in parking surfaces.

Over time, library building envelopes age and break down. Heavy rainfall and melting snow inundate drainage systems, which can't contain torrents of excess water. And wherever there's an opening, no matter how tiny, water enters.

25.1 The human factor

While we're loath to admit it, human error leads to numerous moisture problems. Consider your healthy habit of drinking mineral water while inputting data in your Tech Services Department. Good for you: that insipid, lukewarm liquid is so much cleaner than coffee. But as you reach across your desk to grab a pencil, you knock over the plastic bottle, which pours ion-rich H_2O into your keyboard. So much for your keyboard. If you were working on a new laptop crammed with bells and whistles, so much for the laptop, bells, whistles, and possibly the data that the laptop contained. If you haven't backed up the data, you face serious inconvenience as well as embarrassment.

Another frequent error is leaving a window or skylight open during a downpour. You won't find many descriptions of such obvious mistakes in official maintenance reports. Any janitor, however, can tell stories about the Manager of Children's Services who thought that she was letting the sunshine in, but then she went home and it started to rain and a little brook babbled its way across the Storytime area and into Adult Services, where it formed a little lake which became a little ocean, which the Municipal Property Manager didn't like at all. In fact, when he saw the little ocean, he used a number of words that do not appear in any Newbery or Caldecott winner. At least, not yet.

25.2 A river runs through it

Less common than human error, riverine floods can threaten many libraries, especially those located near big rivers. Most cities have grown along or near large bodies of water, some of which will probably flood their surrounding areas in future. Vancouver residents

Robertson on Library Security and Disaster Planning

worry about the Fraser River, which can submerge thousands of acres during a heavy spring runoff. Winnipeggers are equally concerned about the Red River, which flows near densely populated neighborhoods. Name your city or town, and you'll find that a river runs through it, a lake sits beside it, or a reservoir stands near it. Coastal cities can be threatened from several directions, as water flows through them into the sea.

25.3 Planning, lack of

Many libraries have no procedures to deal with water damage. Your disaster plan may contain fire control measures and evacuation instructions; it may also include a list of emergency telephone numbers. But so what? How useful will that plan be, shortly after you discover that the sprinkler system has accidentally soaked your reference and fiction collections? You might spend time calling various persons and organizations that could help you if they had the time, personnel, and equipment; unfortunately they're too busy to respond to your call for at least 3 days. So much for your collections. Unfortunately, many librarians recognize the weakness of their plans only after the loss of valuable assets. A flood can be the most effective as well as the most brutal lesson in the need for disaster planning.

Poor planning often leads to inadequate clean-up after a flood. The water that you remove from a floor might be only a fraction of the total amount of moisture that has seeped into your building. Your wall cavities and subfloors might contain gallons of water that could soak electrical circuitry and contribute to mold growth. Your collections might be safe, but a week after the flood, staff members and patrons notice an unpleasant odor in the stacks. It smells like mildew. Only when a moisture control specialist opens the affected wall cavity will you find the source of the odor: a pool of water filled with dust, dead insects, and myriad microorganisms—a veritable recipe for a stink.

Since water damage can intensify after the initial flood or leak, inadequate clean-up can lead to much higher repair bills. A prompt, comprehensive response to water ingress is essential if libraries are to avoid its long-term consequences.

25.4 Know thy building

Effective planning and preparations begin with knowledge of your building. Has it flooded in the past? Have its sprinklers activated without appropriate cause? In what condition are the plumbing and roof? Examine your building plans, which you'll need to identify possible areas where water might collect. Skim your library's maintenance reports for any record of repairs to the plumbing, roof, and building envelope.

Also note any proximity risks that might affect your building: the charming fountain fed by nineteenth-century plumbing, the stream that meanders along the edge of the site, perhaps a Great Lake or two. What appears stable and safe now might not remain that way. To recognize potential threats from any source is to begin the process of mitigating them.

A preventative maintenance program can obviate many water ingress problems. All staff members can involve themselves by reporting any water that appears in the wrong place. That stain on the ceiling might be nothing more than a harmless chemical change on the surface of the tiles. Or it could be the result of a tiny leak in a pipe. Have your maintenance department check it before it becomes a torrent that soaks your IT operations room. What about that pool of water under the sink in the staff room, that musty smell in the storage area, and the stains on the concrete wall in the foyer? Don't hesitate to report these items to your maintenance supervisor, and ensure that he or she makes the necessary inspection and, if necessary, repairs.

25.5 Bibliotriage

Naturally, librarians are most concerned with the conservation of damp books. Collection managers sometimes include lists of specific volumes that must be dried out and repaired before any others. This "bibliotriage" can be the most detailed part of a plan. Unfortunately, after damp books have been sorted into high- and low-priority categories, there might be few realistic conservational procedures that can be carried out promptly. Guesswork and disappointments are likely. You note that your soggy Physics collection contains numerous unique and valuable volumes. If you turn to your list of emergency telephone numbers, you'll find the name of a firm in Texas that repairs moisture-damaged books. You phone the firm. The owner tells you that he doesn't work in Canada any longer, because he has too much business in his home state. He's not sure why his name and number appear on your list; nobody told him about it. Meanwhile, mold and discoloration spread across the title pages of your priceless Einstein first editions.

To guarantee that vendors of any kind will respond to your call for emergency service, develop strategic alliances with them. Contact them before trouble occurs and find out whether they're prepared to travel to your site and provide their respective services, be they specialists in conservation, moisture removal, plumbing and electrical repairs, carpentry, or IT systems recovery. Ask your vendors about practical matters. Do they have plans to continue their operations when a regional disaster disrupts local transportation routes and communications? What about the cost of postdisaster service? A strategic alliance should cover all of these issues. Ideally, the terms of your alliances are specified in a contract or letter of intention.

25.6 Remember Noah

While leaks and floods rarely threaten human life in Canadian libraries, there are risks of slip-and-fall injuries in public and staff areas where water collects in pools. Moreover, water and electrical circuitry should not mix. If there's any threat of electrocution, it's wise to turn off the power to your building and wait for a professional inspection of any electrical components that have been drenched. You should include these safety measures in your emergency response plan.

You may never need to worry about an earthquake, and the chances of your library's destruction by fire may be remote. But a leak or flood is likely at some point in future, and the larger the library, the greater the risk from water ingress. Some might accuse you of over-caution. Ignore them. Remember Noah and his critics, and prepare to stay dry.

25.7 Your strategic alliance with a moisture control vendor: What it involves

- Identification of a moisture control vendor in your community who can travel quickly to your site. Remember that janitorial firms are often unqualified to deal with serious moisture problems. Choose a firm that specializes in water ingress and its consequences, and that has the necessary tools to dry out your facility and assets: dehumidifiers, fans, freezers, and cleaning agents.
- Identification of professional conservators for all library media including paper and bindings, photographs and maps, videos, sound recordings, and digital resources.
- Identification of an IT vendor to repair or replace damaged hardware and other resources.
- An emergency contact list that contains the vendor's name, address, telephone number(s), e-mail address, and 24-h contact name(s).
- A contract or letter of agreement that states the vendor's willingness to make the best effort to travel to your library in the event of water damage. Costs of the vendor's services can be included. The vendor might require an inspection of all areas of your library before problems arise, to ensure that he or she has the necessary supplies and equipment to handle moisture-related problems at your site.

Postponing Alexandria: Dealing with catastrophes and disasters in your library

26

Most ancient and medieval libraries have disappeared altogether. Library history reminds us that inevitably, our cherished institutions must deteriorate and follow the Library of Alexandria into the dust. War and revolution, floods, fires, storms, pandemics, and earthquakes have destroyed libraries and other information resources for millennia. We are no closer to eliminating these threats than we were during the earliest stages of civilization.

The remaining bits and pieces of ruined libraries—shattered columns and stone husks of buildings—tantalize archeologists, while scraps of manuscripts inspire codicologists and other specialists to develop new translations and interpretations. For a time, all is not lost—until our own culture decays and humanity loses the last remnants of earlier cultures.

26.1 Writing on the wall

Regional catastrophes claim the attention of our news media. It seems that if reporters and commentators aren't covering a hurricane, terrorist act, or pandemic, then they're predicting one. Smaller events—referred to as disasters or emergencies—receive less attention, but remain newsworthy.

Librarians complain that when a catastrophe or disaster wrecks the infrastructure of a major city, reporters seldom pay attention to the plight of local libraries. For example, the libraries of New Orleans and parts of Florida sustained serious damage during the hurricanes of 2005, but it was the local hospitals, retail areas, and other public spaces that attracted the TV crews.

26.2 Risks and their effects

Fortunately, librarians have developed techniques to mitigate risks to their operations. Wiping out risks completely is impossible, but we can lessen their effects to the point that they no longer pose so great a threat; they may cause us grief, but we might not suffer as much.

The first step in dealing with future catastrophes and disasters is to analyze their possible effects. The risks that threaten your library are often so easy to spot that you might miss them altogether, but now is the time to belabor the obvious. A library beside a river may be situated on a flood plain. A library in an earthquake zone might be in danger of collapsing because of ground motion. Where it rains and snows heavily, libraries leak. If you store your servers in a damp basement, your data are in peril.

Robertson on Library Security and Disaster Planning

Why do we tend to ignore the obvious? Because like the dog down the block, it hasn't bitten us yet, and if we ignore it, it'll probably ignore us. This is the sort of thinking that allows us to live with the abundance of risk in our lives. In fact, we focus on things that are less likely to hurt us so that we can avoid thinking about more frightening threats to our personal safety. For instance, we may worry about another 9/11, but we're far more likely to be injured in a car accident than die in a terrorist attack.

In the library, we may set up committees to discuss slip-and-fall accidents in the parking lot, but a greater threat could be the roof leak that's drenching the electrical circuitry in the server room.

26.3 Mitigation measures

Once you've identified the risks to your library, mitigate them. You can't stop the wind from blowing or the earth from shaking, but you can prepare your buildings, collections, and staff to survive the worst. You can mitigate any risk as long as you identify it before it turns into an active threat.

Mitigation measures involve preparing personnel and assets for all risks noted in the risk analysis. Every day, leaky roofs and plumbing cause enormous damage to library assets. You can mitigate this risk by inspecting roofs and plumbing regularly. If a small crack appears on a pipe or roof section, repair it without delay. Keep library drains clear. It sounds simple. That's because it is.

In an earthquake zone such as southwestern British Columbia, library employees should be trained not to run outside during the shaking. If they remain inside and "duck, cover, and hold" as current disaster preparedness guidelines recommend, they are far less likely to be hit by falling masonry, the cause of many earthquake fatalities.

Librarians can mitigate fire risks by ensuring that electrical circuitry is in good condition, and that employees leave the building when the fire alarm sounds. (How obvious! But in Canada, many libraries no longer bother with fire drills. If they hear the alarm, librarians might assume that a prankster is at work.)

Data loss? Back up essential data from your technical services department and store the backup media in a secure off-site location. Medical emergencies involving elderly patrons? Train employees to recognize signs and symptoms of heart attacks and strokes, and encourage them to learn first aid. Power outages? Store flashlights in all departments.

Most mitigation measures are easy enough to arrange. The challenge is to motivate ourselves to take responsibility for their implementation.

26.4 Response measures for life safety

The appropriate response to an emergency, disaster, or catastrophe is to protect life and limb, not data, carpets, and shelving. Life safety procedures can be laid out in a brochure small enough to fit into a wallet or purse. Brochure instructions should cover all items listed in your library's list of risks.

It's wise not to stuff big binders with emergency guidelines. Few of us bother to read any policy document longer than a few pages, so the compilation of hundreds of pages is usually a waste of time. Prefer the small brochure with advice in point form. Hand out your library's brochure with the pay stubs, or distribute it during employee training and orientation sessions.

To motivate librarians to take life safety measures seriously, ask them to imagine themselves in different kinds of danger. Get them to think not only about themselves, but also about their families. A librarian with these personal concerns will be more inclined to read that brochure and take the necessary first aid and disaster response training than one who assumes that the mayor will quickly come to the rescue. We become more self-reliant when we realize that most civil authorities will be unable to respond to our calls for help during a disaster, and that we'll probably be on our own.

26.5 Resumption measures

After an event that disrupts library operations, you should resume operations as soon as possible, by developing measures to decrease downtime and to facilitate ongoing activities. These may include damage assessment guidelines, collection preservation strategies, service resumption processes, arrangements for alternative sites, and public relations procedures to reassure patrons that their library will survive.

Librarians should organize strategic alliances with key suppliers. Most libraries are unable to resume or continue operations after a catastrophe or disaster without external assistance. Even a minor emergency such as a small flood can shut down a library for days, unless a qualified moisture control vendor is available to repair the damage. Many moisture control vendors are also qualified to handle damage from fires, toxic spills, and high winds. These vendors offer invaluable services after different kinds of events.

A strategic alliance shouldn't cost your library anything until you require the vendor's services. You can prearrange prices, site access, equipment delivery, and clean-up schedules to meet your library's needs. Without an alliance, you might wait days for a vendor to start drying out your collections. It might be too late for many drenched items, which you might have to discard. Particularly after a region-wide catastrophe, a strategic alliance has saved not only a collection, but also an entire library system.

26.6 You've already made a start

With catastrophes hitting the headlines so frequently, you and your colleagues are probably interested in developing plans to deal with them. You've made a start by asking your local fire department to inspect your branches annually, establishing a conservation program for fragile and rare items, and storing your vital data off-site.

You might already have strategic alliances with your regular vendors. It makes sense to build on these practices and arrive at a comprehensive risk management plan. You might think that you're merely postponing the inevitable. And so you are. But your patrons are still eager to browse in your stacks, to ask reference questions, and to attend Children's Storytime. Plan to postpone the inevitable for their sake, as well as your own.

Pandemic perspective: How an outbreak could affect libraries

27

It might *not* happen. Despite recent warnings from scientists, public health authorities and politicians, an influenza pandemic is not inevitable. If one occurs, its effects could be deadly, or it might cause nothing more than a sore throat and a runny nose. We cannot predict the future with total accuracy.

Still, it's a useful exercise to consider what might happen if avian flu spread to human beings, and large numbers fell ill. Many scientists believe that the outbreak of a major pandemic could begin in Asia, and spread quickly in all directions until global morbidity was high. The transmission from Asia to other continents might take less than a fortnight, or more than a year.

International headlines will cause alarm. Politicians will urge us to stay calm. Pundits will wonder why we're not better prepared. Hospital laboratories identify the first Canadian cases as the disease strikes a seniors' residence in Toronto or a high school in Vancouver, and panic takes hold across the country. At this point, librarians could face unpleasant decisions and situations that arise from the public's demands for more preventative measures. Where are the vaccines? What are the measures that will save us?

Even if researchers develop treatments for the disease, it's unlikely that the pharmaceutical industry could produce enough for all of us. Moreover, there's no guarantee that any treatment would be effective. The inefficacy of antiflu drugs would remind us that our science isn't omniscient, and that most of our major discoveries are the result of decades of research, of trial and error. Ironically, a potent remedy to influenza might not become available until the pandemic had wound down, a year or so after it had broken out in a village in Vietnam or Northern China.

27.1 Closed for business

Closure of public and high-traffic facilities would be one measure that authorities might demand. The implications of an indefinite shutdown for libraries and their staffs could be severe. Total closure would require libraries to lock their doors and to cancel services at library sites until further notice. Staff members—including senior managers—would not be permitted to remain at or visit their workplaces without the permission of local police and public health officials.

For some public libraries, there might be security risks during prolonged closures. Serious vandalism, break-ins, and burglaries could become more frequent in some communities. With staff members away from their sites and limited inspections of library facilities, problems would arise from the neglect of basic maintenance and sanitation. If a pipe bursts in the basement, nobody would discover the flood for weeks.

If the sprinkler system malfunctioned on the third stack level, thousands of books might be drenched. Many of them might have dried out by the time we returned to our jobs.

Meanwhile, our facilities could become dirty. With the heating, ventilation, and air conditioning systems working at minimal rates, or ceasing to work altogether, the dust would settle on thousands of shelves and millions of books. There would be no maintenance workers to dust and vacuum collection areas. The filth on library windows from a mixture of condensation and dust would block out daylight and darken our work areas. A musty smell could pervade the building; eventually every curtain and rug would need cleaning.

But some library inhabitants would run free. Rats and mice could invade the staff room and the children's section. Silverfish would move from the basement and washrooms into the stacks and the Technical Services Department. A pigeon could fly in through a broken window on the ground floor and leave its mark of disrespect on a former mayor's bronze bust in the foyer.

27.2 Parking lot returns

Taps drip. Food left in the staffroom refrigerator rots. Mail piles up. Since the civil authorities have ordered the library to lock its book drop, frustrated patrons shove dozens of books and other items through the mail slot; others are left in little piles by the front door. Some of the latter disappear forever. Months later, after the pandemic has subsided and the library has reopened, a regular patron returns a box of books that she discovered in the parking lot. Apparently the borrower who left them there didn't want to be late in returning them, but was too nervous to approach the front door of a public institution during the pandemic.

Does this scenario sound far-fetched? When you consider the effects of even short-term neglect of a building, it's not. Librarians have seen such things after a long weekend or a Christmas break; what they would see in a library that has been closed for several months could be much worse.

Layoffs would occur in many libraries. These could be prolonged and financially inconvenient for most staff members. Still, with daily broadcasts concerning fatalities and the continuing spread of the disease, few staff members would want to risk a shift at the circulation counter or reference desk. Librarians might want to continue some services electronically. For example, limited readers' advisory and reference services could be available through websites and e-mail. Telephone service could continue in some communities. The success of such operational continuity would depend on the sophistication of a library's electronic resources, and on the librarians' ability to work with restricted access to physical collections at their library sites.

27.3 Clean teams

Libraries might also consider the activation of "clean teams" to continue library operations. The origin of such teams is centuries old. In Boccaccio's *Decameron*, written between 1350 and 1353, we find the notion of isolating a select group of people so that they can survive and keep busy during an outbreak of bubonic plague. As long as all members of the group are free of infection, they can work and play together safely. In modern instances such as the Toronto outbreak of SARS in 2003, clean teams from various organizations holed up in Ontario's cottage country and continued working with laptops.

With wireless technology, clean team operations are even easier to organize and maintain. Team members can perform many if not all of their usual duties on BlackBerries and other handheld devices. Of course, the wireless "solution" is not appropriate for some public library activities. It would be impracticable to offer a children's interactive Storytime through a network of BlackBerries, and programs involving puppets are out of the question. But librarians could adapt various technical service processes for temporary wireless continuation, and library administrators could meet and make decisions as easily through a wireless system as they do through conference calling. The advantage of wireless communications is that they allow clean teams to function without team members' traveling to meet each other. When team members don't meet, the risk of infection decreases even further.

Currently, to control the spread of disease, public health officials urge us to wash our hands often. They assume that we touch our faces frequently, particularly our noses, eyes, and lips, and thus infect ourselves with colds and influenza. To wash our hands with soap and hot water is one way to stay healthy. But such measures are less effective if the disease is spread through airborne transmission, a possibility that deeply concerns the medical community. If a careless sneeze can lead to the transmission of deadly influenza, hand washing alone will not stop the spread of the disease. Even with an aggressive national immunization program, we might be faced with high morbidity, especially in crowded urban areas.

27.4 Maintaining health

The most basic way of protecting oneself is to maintain good general health. This sounds obvious, but consider how easy—and comfortable—it is to maintain bad habits such as smoking, excessive eating, and avoidance of exercise. Doctors believe that we can prevent much illness through simple, regular exercise such as walking and swimming. More advanced exercise programs involving aerobics and yoga can stimulate our immune systems even more. Along with a sensible diet and a reduction of stress, exercise can protect us from many different forms of disease, including pandemic influenza.

Many libraries urge staff members to get a flu shot every year. This measure can be effective against the influenza outbreaks that occur during flu season, but it will not necessarily protect us from a form of flu that has mutated from the avian variety. In a pandemic, we would be forced to fall back on emergency response measures such as closure of public institutions and distribution of available anti-influenza treatments. Travel, even between local neighborhoods, could be discouraged or even restricted for a temporary period.

Fortunately, pandemics abate in time, and even the most virulent strain of influenza is unlikely to wreak havoc for more than a few months. There's something else that we should all remember about popular panics. They don't always lead to actual disasters. We tend to forget those terrible possibilities that never became reality, such as the spread of Killer Bees.

And with any luck, some day you can file this article with your unused Y2K emergency plan, and return to your Boccaccio.

Disaster? No plan? What a library director should do

28

"I always pass on good advice," wrote Oscar Wilde. "It is the only thing to do with it. It is never of use to oneself." Since you became the director of your library, you've agreed wholeheartedly. When a colleague suggested that your library needed a disaster management plan, you sent e-mails to various department heads advising them to form a disaster management committee. They did. They met on several occasions, drank coffee, and talked. Then they advised you that a disaster management plan would be a splendid idea.

So you hired a consultant to draft a white paper entitled "Prelude to Recovery: Protecting Information in the Twenty-first Century." The consultant charged a substantial amount to advise you that your library needed a plan for what to do if a disaster struck your library. You accepted the white paper, then filed it and returned to managing your library and making sure that good advice flowed through your office like champagne at a wedding.

Just before dawn this morning, however, an electrical fault in one of your new branches started a fire. The branch was engulfed in flames by the time the Fire Department arrived. You received a frantic call from a board member who'd heard the news on the radio.

"What are you going to do now?" said the board member. Notice that it's not a matter of what the board will do, or the library staff, or the civil authorities will do. It's what *you* will do. You have just lost a branch; you are the director; and if you are honest, you're in shock and don't know what to do.

You dig up a copy of that white paper. In these circumstances, it's useless. You call the director of a college library in Winnipeg: you and he attended library school together. Has he any advice for you? No. He has passed on all of the advice he'd received lately, although he can give you some great anecdotes. But you need help, not gossip.

What should you do? Take heart, then take the following steps. Consider them your *pro tem* disaster management plan.

Step 1: *Find out if anyone was injured—or worse—in the fire*. Enquire not only about the safety of your staff, but also about the firefighters and police officers at the scene. If there were any casualties, be prepared to express your concern and condolences. This sounds too obvious to merit mention, but often managers concentrate on the wrong things after a disaster. You're not heartless, but in shock you might tell a news media representative that while the injuries to one of the firefighters are unfortunate, you've lost a new branch that cost millions and took a year to build. You would regret saying this for the rest of your life, so don't. Tell the media, the board, and your staff that you are saddened by the human cost—which most likely is the truth.

Step 2: *Find out the extent of the damage, and estimate the effects on your library system*. If the branch has been destroyed, you may take it for granted that you will not be able to offer as much service as before. But did that branch contain your system's servers, or a special collection? Did it cater to particular groups such as seniors, physically challenged patrons, or clubs? The data on the servers have been backed up, of course. You can restore the system's catalog and borrowers' records within 24 hours. (If you're reading this before a disaster has struck, and you haven't backed up your vital data, do so now. No further delays. If you have actually lost your data because you haven't backed it up, you and your library are in serious trouble. You must ask staff to resume manual procedures for checkout and processing, and inform your users that the catalog will be unavailable until further notice. Then the technical services staff must begin reconstructing lost data, a time-consuming and expensive task. But you're a sensible and responsible professional, and you've made sure that your data are properly backed up, haven't you? If not, you will, won't you?)

Step 3: *Prepare to manage a crisis, which is defined as a negative media response to a problem in your organization.* In this instance, crisis management involves making yourself available to answer questions about the fire and its implications for your library. Remember that when you speak to the media, you're talking to everyone in your community, including library staff members. You'll begin by expressing regrets about any casualties, and sympathize with their family members. Then you'll mention the damage to your collections—special and otherwise—and make known your determination to restore them. You're happy to report that services to seniors and the Summertime Reading Club will resume tomorrow at a different branch. At this point, you must avoid the error that too many directors make: giving up on an asset. *You will affirm that you might have lost a building, but you haven't lost a branch.* You intend to do everything in your power to rebuild what you've lost. You express the hope that the community will continue its support for the library. To say that you've lost a branch is to allow those higher authorities that control budgets to postpone its reconstruction, or to let your system lose it altogether. They will tell you that economies must be made in hard times, and that funding to replace the building might not be available for years. But if you have let your community know that the branch continues to exist—in spirit, at least—then it is politically more difficult for higher authorities to let you lose the building and the branch that it houses forever. Is this playing politics? Yes. Make sure that you're better at it than those who would allow your library service to decline.

Step 4: *Support your staff.* They will look to you for something more than leadership or direction. They'll want you to show your humane side. They might need frequent reassurance that the library—and their jobs—will survive. Some staff members will be emotionally hurt by the loss of a library building, even if they've never worked there. Loss of a special collection can be heartbreaking for the staff members who developed and conserved it, cataloged it, and showed it off proudly for years. Tell your staff that you empathize with them, that you too are struggling to deal with adverse circumstances. But reinforce the message that the library will survive, and that full service will resume as soon as possible. To staff members worried about layoffs, tell the truth. After a disaster in a public or academic library, staff members might have too much work. Rather than lose their jobs, they might have to deal with even more hectic schedules and overtime. In fact, job loss following a library disaster is uncommon. Severe stress and exhaustion, however, are definite risks. Your Human Resources department must ensure that staff members are not assigned too much overtime, and that library sites maintain the appropriate staff complement. Remember that at least one of your branches will take on services that had been available at the building that you've lost. Staff members from the lost building will probably have lots of work at other branches.

Step 5: *Deputize*. You've been talking about your management team since you took the director's job. Now demonstrate that the team functions well, and that when the library faces difficulties, trusted colleagues can assist you. Rely on them to conduct damage assessments, adjust schedules, and deal with concerned staff members and patrons. Ask your senior deputy to initiate fund-raising to replace the lost building. It's never too early to look for the money for such a cause. Don't take on too many responsibilities yourself, or you'll soon be exhausted. Above all, let the more talented and energetic managers rise to the occasion. Remember that you can't do everything.

Step 6: *Normalize*. It takes time for any organization to resume normal operations after a disaster. In fact, many staff members become disorientated and lose their sense of "normal" after an event that results in serious loss. Working at the library might feel odd. Patrons keep asking whether the burnt branch will be rebuilt. Despite your frequent denials, rumors about layoffs might spread. At least one staff member takes stress leave; others are considering it. In these circumstances, your message must be clear. You and your deputies will do everything possible to ensure that the branch will be rebuilt, and that staff members will keep their jobs. Ask staff members to get in touch with you if they have any pressing questions about the system's recovery, or if they can recommend effective ways to resume normal operations more quickly. Encourage staff members to speak openly about how they feel. This sort of "verbal therapy," as emergency management specialists call it, can help staff members to obviate doubts and other negative feelings about the library's status. Good morale will depend on staff members' confidence that they can speak out and confirm their roles in the library's recovery. Commend staff members for providing comments and suggestions, thank them for their contributions, and let them know about whatever progress you make regarding fund-raising and resuming normal operations.

Step 7: *Plan for the future*. Even a serious loss of assets should not stop you and your management colleagues from making short-, medium-, and long-term plans for your library. You will be tempted to delay projects for extended periods, until you know how the board and general public will respond to the loss. But the sooner you can return to normal or near-normal scheduling of operations, the better for your staff and patrons. Nobody likes to get caught in a postdisaster rut of postponements and uncertainty. Moreover, your board and other authorities will be impressed and relieved to see the library making progress despite disaster-related disruptions.

You can, of course, simply pass this advice on to somebody else. Feel free to do so if you've already developed and tested a comprehensive disaster management plan. Otherwise you'd be prudent to make a note of the 7 steps described in this chapter, and prepare to give the performance of your career to the news-hungry media and a perturbed board. With luck you'll win an Oscar. In the worst circumstances, no other result will be acceptable.

Out! A guide to emergency evacuations

29

Voices in Jimmy's head tell him that libraries are evil. One voice insists that librarians are wicked and must be stopped.

"Blow up the library," says the voice. "Stop the librarians from hurting good people."

Jimmy sits at a library Internet station and learns how to construct a bomb. One website demonstrates how even the simplest and least expensive detonator can trigger a deadly explosion. Inspired, Jimmy builds a bomb that he believes will flatten his local library. It happens to be your branch.

A few minutes ago, Jimmy walked past your reference desk with an old briefcase tucked under his arm. He disappeared into the stacks. Shortly thereafter, he dashed out the main entrance empty-handed.

The telephone on the reference desk rings.

"I planted a bomb in the library," says an angry male voice. "It's going to blow up and kill you. You are disgusting, and I am going to stop you."

What now?

29.1 Procedures

Your library might have bomb threat response procedures buried somewhere in a manual. These should contain standard advice including directions to cooperate with the caller on the line, ask for the location of the bomb, and try to find out when it will explode. Ask why the bomb has been planted. Pay attention to the characteristics of the caller's voice, which will reveal the gender of the caller, and sometimes a particular accent or dialect. The caller might sound intoxicated. He or she might also sound familiar, and in this case, you suspect that Jimmy is on the line.

Where is the caller now? Can you hear any background noise, such as traffic or construction? Can you hear music or other voices? You might have only a few seconds to ask questions and to note any voice characteristics and background noise. The caller could hang up at any time. Obviously you should not put him on hold, interrupt him, or make threats in return. Unfortunately, some people make these mistakes.

The call ends abruptly. The bomb threat response procedures might tell you to contact your library's security department immediately. You should call 911 as well, and answer the operator's questions clearly. Do not touch any suspicious item such as a parcel, briefcase, or knapsack that has been abandoned in a washroom, stack area, or other public space. An emergency team—the police—will be dispatched in a minute or two; in fact, you can already hear sirens in the distance.

29.2 Time to go

And now it's time to evacuate the library. The procedures might tell you to trigger the fire alarm and leave the building. Or they might advise you to tell—calmly and firmly—all library employees and patrons to leave as soon as possible by the nearest exit, and not to return until you indicate that it is safe to do so. In some jurisdictions, however, the police prefer you to remain in your building until they arrive, so that you can point out any mysterious packages or other suspicious items. But how comfortable will you be while waiting for the police to arrive? What if the bomb were to explode before they arrived on the scene? How happy would library employees be about remaining in the building with a bomb? With all due respect to the local police, many librarians will not hesitate to evacuate their buildings.

But some people might be slow to move toward the exits. Some might not want to leave the library. They—including staff members as well as patrons—might not believe that it is necessary. Or their curiosity could inspire them to waste time asking questions. Is it really necessary to leave? Is this a drill or an actual emergency? Should we turn off our computers?

Moreover, people do not enjoy evacuating quickly, without their coats, into a snowstorm, freezing temperatures, or heavy rain. But would they prefer to stay indoors to feel the blast of a bomb or the effects of something equally dangerous?

Hence most North American workplaces are obliged to hold regular—often annual—evacuation drills, also known as fire drills. Such drills are appropriate responses to a number of different threats, including toxic spills and hazards such as smoke and fumes emanating from neighboring sites. Drills are not as simple and straightforward as you might think. Until you organize a drill for library employees, you might assume that they know automatically where the nearest exits are, and where to go after they arrive outside. Frequently they do not. They might never have read your library's emergency manual. If they are new to your workplace, they might not be familiar with various exits.

29.3 Intelligent response

The goal of any drill is to get people to respond intelligently to an alarm. There should be no question about what to do when the alarm sounds. People should not dawdle, or try to retrieve lunch bags from the staffroom fridge, or Google "fire alarm" at the reference desk. As for patrons, the library employee in charge of staging and leading the evacuation—most commonly the designated fire warden—might advise them, over the public address system, to leave the premises immediately. Some wardens move briskly through their library's public areas, telling patrons to leave at once. In most jurisdictions, that is all that a warden is legally required to do for patrons. A warden is not forced to remain with a patron who refuses to evacuate, nor should a warden attempt to move any patron physically. If a patron requests assistance, a warden may offer to help him or her. Sometimes elderly patrons and those who are physically challenged might need help, and a warden can offer them a hand. The best result is that *everybody* who was in the library has left quickly and arrived at a safe place.

There are impediments to fast evacuation. You might not like to admit that your branch's fire exit door is difficult to open. Or perhaps a carousel stands in front of it, or a book sale cart. Your fire warden should ensure that the door opens easily. If the lock sticks or the hinges creak, you should ask the custodial staff to perform the necessary maintenance. And while they are in the branch, ask them to replace the light bulb that burned out a year ago in the exit sign.

Another impediment is inappropriate footwear. Ideally, library employees don flats for workplace activities. While an attractive addition to evening garb, high heels can hinder fast movement and should be discouraged. For employees who must evacuate down external fire escape stairs that are slippery, high heels of any style can make the descent even more nerve-wracking.

29.4 Burberry versus life

Of course, people do not like to leave behind personal belongings including purses, coats, and laptops. Your library's safety orientation program should encourage them to consider their priorities. They may value their wallets, and wince at the thought of having to apply for replacement driver's licenses and credit cards—but is it worth risking their lives for such things? That Burberry coat was a Christmas gift, and perfect for rainy days, but it can be replaced. You, however, cannot be replaced so easily, and neither can the owner of that splendid new laptop. Life safety is paramount; before it, everything else is negligible. But people need to be reminded of this through effective orientation programs. Having evacuated, library employees should assemble at a safe gathering site, an area some distance from the building but not too far away: perhaps no more than a couple of blocks. Your safe gathering site should not be affected by the smoke and fumes of a fire, the vapor from a toxic spill, or the blast of an explosion. It must be large enough to accommodate your entire staff. If possible, employees should not have to cross a street to reach it. And once there, the fire warden must make sure that all employees have arrived at the site. For staffs comprising more than 15 people, the warden might need a staff list to call the roll and to confirm that everyone is present. A person missing from a smaller staff will be conspicuous by his or her absence. In a real emergency, anyone missing—and any patron who has refused to leave the library—should be reported to the first police or firefighters at the scene.

29.5 Real bombs are rare

Bomb threats require thorough investigation. If you have evacuated your branch, the police will ask you for details regarding the layout of the branch, and whether it contains places where somebody might be more likely to plant a bomb. They will ask you if you have any idea about who might do so. They will begin their inspection, sometimes with the aid of bomb-sniffing dogs. The planting of real bombs in North American public places is rare, but the police take every bomb threat seriously. If the bomb in your branch appears capable of exploding, a specialized bomb squad will remove it to a remote or bombproof location and detonate it safely. Once they have removed the bomb, the police will allow you to return to your workplace. Your library's senior management will ask for a full report soon after, and will want reassurance that the threat will not be repeated. The police will try to find the culprit, and fortunately in this case you have given them a lead.

You should keep in mind that different threats will lead to variations in your evacuation technique. A fire in your building will force you to make a straightforward evacuation to a specific safe gathering site. But a fire in a neighboring building could demand different evacuation routes and safe gathering sites. If a tanker truck were to collide with another vehicle and spill harmful chemicals onto the street outside your fire exit, obviously you should not take that exit to avoid the spill. Some fires and toxic spills can lead to employees' walking greater distances to avoid smoke and noxious fumes. Orientation programs should attempt to address these and other possible events.

29.6 Stay put

In seismic zones such as British Columbia and California, the risk of an earthquake will demand counterintuitive measures. During an earthquake it is wise *not* to exit or enter a building, since falling masonry can kill. Library employees may be tempted to evacuate their buildings when the ground begins to shake, but instead they should "duck and cover" to stay clear of falling objects. Dedicated orientation is needed to accustom employees to the idea that they should not run from a building during an earthquake, and even after such orientation, some employees might still dash from their buildings.

Thus evacuating your library involves more than nipping out the exit and having coffee down the block. You must assess prevailing risks to your site and determine which ones should lead to evacuation. You should appoint fire wardens for your building—at least 1 for each floor, and ideally 2 or 3 to cover each other's holidays and absences. You should identify a safe gathering site, and consider an alternative site in case the primary one becomes unsafe or inaccessible. An orientation session for all employees should include an oral presentation, and a drill supervised by the fire wardens. You should repeat that drill at least once a year and not worry about inconveniencing patrons. You are acting in their best interest and for their safety.

29.7 Extinguishers

Some orientation programs include training in the use of fire extinguishers and in the evacuation of injured persons. The fire wardens should take this training, which fire departments in some jurisdictions offer free of charge. But start slowly, to make safe evacuation and response to various threats a part of your library's standard operating procedure. Trying to accomplish too much in too short a time can lead to employees' indifference to essential points of any safety program in the long term.

As for Jimmy's bomb, it turns out to be a broken digital clock taped to a bag of firecrackers. Jimmy tells the police officers who knock on his door that a voice has told him to destroy your library. He is taken into custody and found to be a danger to himself and the public. But in his care facility, a mental health worker reads him stories and eventually accompanies him on visits to a library where he feels secure and welcome. He enjoys reading coffee table books about nature and foreign places. He tells the mental health worker that libraries are not such bad places after all.

Nevertheless, library employees should know how to leave their workplaces safely, just in case. Poor Jimmy is not the only person who hears voices.

Flood protection: Avoiding the wrong mix

30

Water and books don't mix. Nor do water and any other library asset. Hence librarians are wise to note that water can make an entrance in many different ways. It can seep into a library basement through a leaky pipe, or drip onto stacks from a crack in the roof. During a rainstorm, it can pour through a window or skylight that somebody forgot to close. Or it can surge into a library during a flood from a river or other overflowing natural source, and cause extraordinary damage.

Generations of librarians began their study of library history with a consideration of the Library of Alexandria and its destruction by fire, which has always been a serious risk. Currently, however, water cause of far more losses in North American libraries and other information institutions. Ironically, many librarians rely on sprinkler systems to protect themselves and their collections from fire, only to discover that sprinklers can wreak havoc when they trigger accidentally, owing to a mechanical fault. Thus what is intended to protect our assets can also destroy them. And even if we are constantly vigilant, there are occasions when, despite the best intentions and preparations, we cannot prevent significant losses. Along with large earthquakes and high winds, flooding by rivers and other large bodies of water can present librarians with profoundly challenging circumstances.

In 2013, Calgary's huge flood made librarians across North America acknowledge the possibility of a similar catastrophe in their own localities. Calgary will need years to recover, both physically and psychologically. Government sources and the insurance industry have predicted losses of more than $5 billion. A noteworthy aspect of the media coverage of Calgary's tragedy is that it did not travel far. While Canadian media concentrated on the plight of Calgary's citizens for weeks, American, Asian, and European coverage was cursory and short-lived, if not negligible.

"In fact, monster floods are increasingly frequent," says Paul, an environmental engineer who has worked in Toronto and Vancouver. "With climate change, we can expect flooding in urban areas in the foreseeable future. With big floods occurring more often, the media will focus on those that result in high casualty figures, with hundreds or thousands of fatalities. The good news in Calgary is that while physical damage was significant, few people died. You'll hear about floods in Asia that kill thousands."

Paul does not dismiss the possibility of floods in North America even more destructive than the one that struck Calgary. He believes that, since so many of our communities have developed near bodies of water, we must accept that our vital institutions and infrastructure are at risk. But he insists that there are effective methods to protect our libraries.

"Libraries are vulnerable," he says, "but there are good ways to prevent losses, or to reduce them in a situation involving a *force majeure*, that is, a set of circumstances that are simply too great, too devastating for us to ward off. You can prevent water ingress from a leaky roof by inspecting the roof and making sure that it doesn't leak. But some floods are so severe that you can't stop them. They overflow dykes and submerge entire urban areas. In such cases, libraries might be damaged, but if they have mitigated flood risks properly, they will not lose nearly as much as other institutions and they will be back in operation a lot faster."

How can you prepare your library for all forms of water ingress and flooding, from the smallest plumbing leak to a major flood that submerges entire buildings? There are 6 essential steps of preparation.

30.1 Step 1: Identify the risks and their sources

Take nothing for granted. Consider the plumbing—including sprinkler systems—and its overall condition. Have there been leaks in the past? And what about roofing, gutters, and drainage systems? Has there been any evidence of leaks?

"Many cases of water damage start small," says Mitchell, a Vancouver carpenter and plumbing contractor. "Sometimes you'll see a small brown stain on a ceiling or wall. It could be from condensation, but it might also be due to a tiny leak in a pipe—a leak that could become larger over time. Such stains should be examined carefully to make sure that they don't get worse in future. Often a small leak will lead to a rupture, and then you'll see all sorts of interior damage—and for librarians that means wet books and other items. Fact is, that kind of disaster is preventable, as long as librarians tell their property managers the location and extent of the problem when they first detect it."

Outside library buildings, other mechanical and technological risks prevail. Water mains can burst under streets. Sewers can back up. Dams, dykes, and reservoirs can fail. Meanwhile, the old building next door to your branch has been poorly maintained. It is not only a firetrap, but also a potential source of uncontrolled water, which could flow from a burst pipe into the street, and end up flooding your basement and main floor.

Most North American cities and towns are located near bodies of water such as lakes and rivers. Coastal cities must contend with oceans and their high tides, storm surges, and tsunamis. Bodies of water might not flood often, but when they do, a catastrophe such as Calgary's is possible. Consider the history of those bodies of water in or near your community. Even if there are no floods on record, do not discount the possibility that the gentle stream that flows so scenically near your site cannot break its banks and damage your assets. Many of the world's largest rivers formed from gentle, scenic streams. All bodies of water should be noted for your risk management purposes.

30.2 Step 2: Mitigate the risks as much as possible

You cannot stop a tsunami or heavy rains, but you can make sure that your roof is in good condition, and that your drains will function properly during heavy precipitation. And—a key point during a flood that causes a power outage—review your data backup procedures. Do not assume that the systems librarians have taken care of everything. Ask them how your library's vital data are backed up, and how often. Ask them where the backup media are stored, and whether those media are tested to ensure that they are not deteriorating. Finally, ask them whether those media are accessible after a disaster.

"These days, the Cloud is used to back up data," says Cheryl, a corporate librarian in Vancouver. "That's an acceptable solution, as long as you're comfortable with current Cloud security protocols. Many special librarians and records managers aren't. There are too many concerns about data theft and misuse. For vital, mission-critical data, old-fashioned digital media are often preferred to the Cloud. As long as those media are available after a big flood, your library will be fine."

Another mitigatory measure is to avoid storing valuable collections and other materials in basements and top floors. These levels are frequently the first to sustain water damage during severe weather and floods. Such a measure may not be feasible in libraries that are short of space, but in many cases, the most valuable items can be removed from risky areas and stored elsewhere, with no loss of accessibility or security.

Mitigation is an ongoing process, and you should regularly seek opportunities to decrease the risk of water damage in your library by moving assets away from exposed areas.

30.3 Step 3: Develop and maintain strategic alliances

The first alliances that you form should be with other institutions—including other libraries and information organizations—that might assist you if your site is flooded. Why not exchange data from your technical services department with another library located in a different part of your city or region? (These days, you can exchange data with a library on another continent.)

"You can also make arrangements with other libraries for remote siting of operations," says Darryl, a cataloguer who has written a disaster plan for his library in

Winnipeg. "If your public branch is under water, you might be able to set up offices in a local college library. You needn't rely solely on libraries, however. Local churches, malls, and community centers might be able to offer your staff a temporary workplace in the event of a flood. Before you rent trailers—an expensive solution—ask property managers in your community if they have a place for you to come. You'll be surprised how many people will agree to help out a library."

Another essential alliance for any library is with what in current parlance is called "a moisture control vendor" or MCV. One of the more prominent MCVs operating worldwide is Belfor, whose emergency crews have dealt with thousands of floods, from the largest regional events to those limited to a single room. It is prudent to contact your MCV before a flood or any other disaster occurs, to inform their emergency managers about your particular concerns during any future flood. When the MCV's staffers learn about your site and assets, they will have a better idea about how to respond to rising water in your basement or a drenched collection of reference resources on your mezzanine.

"Phoning an MCV out of the blue after a flood in your library is not the best way to deal with water damage," says Darryl. "Forewarned is forearmed. Besides, arranging an alliance with a vendor before things get wet should not cost you anything. The vendor will appreciate the prior notice, and you'll know whom to call if the river breaks its banks."

Many libraries have loose arrangements with external conservators, who are expected to arrive at a library and to preserve its water-damaged items. Often such arrangements are too loose to be effective. Including a conservator's name and telephone number in your in-house emergency services directory will not guarantee that the conservator will be available to help you after a flood. (She might have her own damp basement to worry about!) It is best to make firm emergency conservation plans with conservators who are prepared to travel to your site as soon as possible after water damage has occurred. Your MCV or your insurers might be able to advise you regarding available conservators in your area.

Remember that while your library's in-house emergency conservation procedures look good on paper, putting them into practice following a flood can be difficult or, in some instances, impossible. Libraries that maintain realistic alliances with external conservators can have a higher level of success in saving damp assets.

30.4 Step 4: Develop flood safety procedures for employees

The purpose of any safety procedures is to keep employees (and others on-site) safe during a flood. You do not require a voluminous safety (or emergency response) plan. Be brief. Your flood safety procedures can include points such as the following:

- Try to determine the source of the water: burst pipe or water main, local river or lake, roof leak, or faulty sprinkler.
- Turn off the water main if the source of water is faulty plumbing.

- Do not walk through water. Avoid contact with floodwater—which might be polluted and toxic—as much as possible.
- Avoid wet electrical outlets, wiring, terminals, and appliances.
- Do not attempt to boot up damp computer hardware.
- Contact the facilities manager as soon as possible.
- Expect a power outage. Keep flashlights handy.
- Do not open the library or begin clean-up until electrical outlets and wiring have been given a safety inspection.
- In the event of a site closure, post signs advising patrons to return items to other branches, etc.

Ideally these procedures should be distributed to all employees. Posting them on a website and assuming that everyone has reviewed them is insufficient. Handing out and reviewing a sheet of instructions during training and testing sessions is more effective.

30.5 Step 5: Develop training and testing programs for employees

Plans and procedures on paper will not protect you unless employees know their contents. Employees can learn what they need to know through training that might include brief presentations by local emergency responders, tabletop exercises, Q&A sessions, and webinars. Testing can include evacuation drills and rehearsals of emergency site closures and the secure removal of the most valuable assets (e.g., the rarest books, manuscripts, and artworks).

"The more you test your procedures, the better prepared you'll be," says Don, a Toronto academic librarian who acts as the fire warden for his building. "Frankly, any paper or website is useless unless people have actually carried out the instructions in a test. You'll find that most staff members will welcome the opportunity to perform a walk-through, and they'll let you know what they think of different procedures. Comments from staff members during tests can be more helpful than the advice of external experts. Staffers know the site better."

30.6 Step 6: Develop a postdisaster or operational resumption plan

Your final step is to plan for post-flood circumstances. Your aim should be to resume operations as soon as possible after your site has passed a safety inspection. Contents of an operational resumption plan can include the following sections:

- Damage assessment procedures
- Emergency communications and PR
- Insurance issues
- Emergency travel procedures
- Site closure and reopening procedures

- Strategic alliance activation procedures
- Clean-up and waste removal activities
- Primary conservation activities
- Staff allocation and emergency succession
- Emergency volunteer policy
- Normalization procedures
- Training, testing, and audit procedures.

The previous 5 steps will provide much of the material that you will need to formulate an operational resumption plan. As with other planning components, it needn't be a large document. Many of the more practical resumption plans are no more than 30 pages. Note that larger the plan, the less likely it is to be read, the more difficult it is to put into effect, and the less useful it will be when water finds its way into your library.

Water mixes with so many things to support us, and to make our lives enjoyable. With soil, to produce crops. With small boats and big ships, to take us all over the world. With hot summer days at the cottage, where we can cool off in the lake. But if lakes start to rise and rivers burst their banks, we need something to protect us and our libraries. Something more than scotch and water, which might be nothing more than a desperate remedy. Something such as a plan. Start one now.

30.7 Generic tabletop exercise: flooding at a large library

At 11:45 p.m., a pipe bursts on the main floor of your building, and ~1500 gallons pour onto the floor and throughout your department's work areas and stacks. Carpets, furniture, books, and wiring are soaked. The flood is not discovered until 6:15 a.m. the next day.

30.7.1 Casualties

Arriving early, one employee slips and falls in a stairwell and sprains his ankle. He lies in the staff entrance for 30 min before he is discovered.

30.7.2 Damage

- Drenched carpets and underlays, which emit an unpleasant odor within 3 h.
- Soaked wiring, which increase the risk of fire and computer hardware malfunction.
- Soaked ceilings and wall cavities, which emit an unpleasant odor within 8 h.
- Wet books—thousands of them.
- Wet artwork, posters, and ornaments mounted on walls.

30.7.3 Crisis potential

- A roving TV reporter hears about the flood while having coffee at a nearby cafeteria. She arrives in the library and looks for an interview. She calls for a camera crew. The resulting story: "Before leaky condos, there were library floods. There still are. Here's what happened to a branch of the local library, and all because of a burst pipe. What are they going to do with those wet computers, those soggy books?"

30.7.4 Your tasks

1. Refer to the library's emergency plan. Make a list of the first steps that you would take to resume business in your water-damaged work areas. (Take 15 min.)
2. Write a schedule with realistic time estimates for the steps that you have listed in 1 (15 min).
3. Your staff members are told to resume business at the remote site. Make lists of:
   - the basic operations that you will be able to conduct within 48 h from the remote site (e.g., communicating with external businesses and key contacts, issuing bulletins, and taking delivery of items)
   - the equipment that you will need to offer the basic services listed above
   - persons and organizations that you must contact as soon as possible regarding library service issues. (This part of the exercise will be limited to 45 min.)

31 Surviving by committee

Faced with earthquakes and other serious risks, the fictional Director of a library in British Columbia tells one of her managers how to use in-house resources to develop effective disaster plans.

I am your Director, and I have a problem.

Yesterday at a meeting of the Library Board, the Chair told me that she "had concerns" about our library's ability to recover from a disaster. Another Board member mentioned that our response to last winter's stormy weather was "less than satisfactory." Somebody else asked about the effects of an earthquake, which is always a concern in a seismic zone such as British Columbia. Then the Chair talked about what she called "basic stuff," such as dealing with fire and water damage. We run a high-profile organization that provides essential services to our community. We should know how to deal with these things.

Here's your task. I want you to convene a disaster planning committee. Your goal will be to present a comprehensive disaster plan to the Board in 6 weeks. You think that's impossible? On the contrary, you could probably finish the job in even less time. Disaster planning committees that produce inadequate plans often have too much time and no clear goal. Remember Parkinson's Law: work expands to fill time. If I gave you 6 months to write a report on library floods, you'd take 6 months. That's human nature.

Begin by inviting key members of each department to join your committee. Tell them that you have a mandate from the Director and Board. Make sure that you get managers from Administration, Human Resources, Facilities, IT Systems, Children's, Outreach and Communications, and Adult Services. Ask the Assistant Director and the Area Managers to get involved. Of course I'm too busy to play an active part, but if you can persuade me to join, do so. The more senior managers you can press into service, the better.

31.1 Agenda

Before the first meeting, circulate an agenda. Tell committee members that they must review our library's current disaster plan and determine what we need to protect staff and assets in the event of a disaster. Do we already have a disaster plan? Perhaps there's a three-ring binder in the library that deals with fire and flooding in the old Main Branch from where we moved in 1981. That's no use now. We need a new plan, and not just on paper.

To cover prevailing risks, we should have a plan that prepares us for what might happen. The more prepared we are, the easier and less costly it will be to recover from an earthquake, fire, or flood. The committee should be able to recommend the best ways to prepare for risks, ideally by mitigating them. A power failure could shut down our operations and wipe out vital data, unless we have an emergency generator, an uninterruptable power supply (UPS), and a comprehensive data backup program. Fire and water damage could disrupt library operations for an extended period, but if we have space reserved at an alternative site—perhaps one of our branches—we could be serving patrons again within 48 h. An earthquake could cause more damage in older structures that have less stable foundations, so we should investigate the condition of all of our buildings—especially our beloved Carnegie Castle, which opened before World War I. Having considered safety requirements and operational priorities during disasters, we will be prepared for a broad range of problems, including damage to facilities and collections, higher absenteeism, and site closures. Ask the committee to make a list of mitigation measures, and be ready to defend each one to the Board.

31.2 Emergency response plan

The next step is to develop an emergency response plan. The local Fire Code and other regulations state that we must have evacuation routes in case of fire, and staff members trained in first aid and CPR. Various governments have issued guidelines for safe response to toxic spills, pandemics, and earthquakes. The committee should develop a plan that includes all the procedures we need to protect life and limb if we are threatened by a physical risk. Remember that a response plan should do nothing more than help us to protect ourselves. It should not cover data backup or repairs to buildings, nor should it try to protect the high-priced art in the boardroom.

We already have parts of an old response plan scattered throughout our offices. Posted beside the elevators are the evacuation route diagrams that nobody looks at; and in the bottom drawer of every Information Desk is a dusty binder labeled "Security," which contains advice on how to deal with violent intruders and bomb threats. The committee should combine these bits and pieces, new and old, in a single document as brief as possible. Then draw up schedules for testing response procedures and for staff training. Be prepared for people to complain about the inconvenience of holding a fire drill, and feel free to remind them that you won't have time to arrange emergency response tests when their building is burning down.

Speaking of training, I must caution you and your fellow committee members about relying on outside expertise: "facilitators" and consultants and other gurus. Often they can provide invaluable advice, and save you time and money. But they won't be around when all Hell breaks loose, so one of your committee's jobs is to wean the library off outside assistance. We must be as self-sufficient as possible in all circumstances, and we cannot depend on consultants indefinitely. Make sure that when you hire them, you have a start date, well-defined deliverables, and a final delivery date laid out in a contract. Do not give a consultant an open-ended assignment that never ends.

31.3 Business resumption plan

For many committees, the greatest challenge arises during business resumption planning, which should follow preparedness and response planning phases. Everyone agrees that we must prepare for disasters and be able to respond to them for life-safety purposes, but when managers from different departments start thinking about resumption priorities, various demons appear. The Children's Services manager assures the committee that his department must take precedence over all others, "for obvious reasons." At that point the IT Systems manager proclaims that her department is the library's nervous system, without which the Children's Services department will be paralyzed. Thus it's only common sense that business resumption starts in the IT operations room. Then the HR manager asks whether she's chopped liver, and the manager from Accounting says yes, she will be, unless the library meets its payroll.

As the committee's Chair, what should you do? First, remind your colleags that business resumption is impossible unless all departments are willing to cooperate. Second, recognize that some library departments must take precedence over others after a disaster. Usually Facilities and IT Systems are top-priority in the first 48 h after a regional disaster such as a flood, winter storm, or earthquake; then Administration and other departments should be scheduled to resume basic services. Finally, recognize that in the first few days after a disaster, not every employee is needed at the library to assist in business resumption activities. Those who are not engaged in reestablishing communications with patrons or assessing damage to collections might find other tasks, or they might simply stay away from the office until they can safely return. Some could work from their residences. You must reinforce the idea that not everyone is necessary on site immediately after a disaster. Often only a skeleton crew is needed for the purposes of security, basic clean-up, and supervising repairs and the delivery of new equipment.

31.4 Be brief

The disaster plan can appear in different forms, under numerous titles. You are expected to present a document to the Board, but make it brief if you want anyone to read it. In fact most boards will be satisfied with an executive summary and a description of your committee's ongoing activities. Keep in mind that no disaster plan is ever finished, since a library's prevailing risk profile changes with new sites, services, and staff. There will be no final and definitive draft. The plan should be audited and updated every year, and different components should be tested regularly to make sure that they're up to date. New staff members need orientation to the emergency response plan, and to any parts of the business resumption plan that directly affects them.

Expect a few searching questions about our ability to recover from winter storms and floods, and the need for emergency conservation of the more valuable collections. Be sure that you mention your investigation of the earthquake risks at older branches. Avoid jargon and bafflegab as much as possible, since obscure technical terminology tends to put people to sleep.

Above all, you and your committee colleagues should use common sense. Without it, the next earthquake or flood could turn us all into chopped liver.

31.5 Six-week agenda

Your disaster planning committee has 6 weeks to prepare a basic disaster plan. Meeting once a week for 2 h, you and your colleagues can cover the basics with this schedule:

Week 1: Reviewing plans and risks

- Make sure that all key departments are represented.
- Review any past disaster plans for your library, including old fire evacuation plans, security programs, and documents concerning recovery from Y2K problems. Determine what's still useful. (Yes, Y2K response procedures can still be useful, especially if they address risks such as power failures and data loss.)
- Agree on current prevailing risks at all library sites. Remember natural risks such as earthquakes and high winds, technological risks such as systems failure and data loss, and human-caused risks such as sabotage and theft.

Week 2: Mitigation and emergency response measures

- Develop a mitigation program focused on prevailing risks. If the library has not done so already, make arrangements to back up essential data, and to comply with local fire codes.
- Develop emergency response measures for life safety. Local fire and police departments will often supply helpful advice and guidelines. Other information sources include your municipal hall and government emergency preparedness agencies.

Week 3: Emergency response training and testing schedules

- Make sure that the library's fire warden systems are effective at all sites. All designated fire wardens should be aware of the proper evacuation procedures.
- Develop schedule for fire and evacuation drills. (By the way, when did you hold your last drill?)
- Investigate outside training resources as required.
- Review your roster of first aid attendants. While there are a minimum number of attendants required by local regulations, you can't have too many on the staff. Discuss the possibility of sponsoring a first aid training course for your library.

Week 4: Business resumption planning components

- Develop procedures for disaster recognition and declaration, a damage assessment checklist, and an emergency contact list with telephone numbers and e-mail addresses.
- Develop a strategy to activate your backup data. Remember that if you can't activate the data, your library could be in trouble.
- Develop a list of strategic alliance partners—the people and organizations that will assist you in getting your library back in operation. Moisture control companies, conservators, IT hardware and software vendors, and alternative site managers are your library's standard strategic alliance partners.
- Develop a tentative schedule for business resumption. Allow yourself 72h to resume your library's basic operations.
- Develop a basic post-disaster communications strategy that enables you to resume contact with patrons and key contacts as soon as possible after a disaster.

Week 5: Business resumption component testing

- Develop tests for resumption plan components including data backup activities and your relocation at an alternative site. You can test most components through tabletop exercises. Your Information Services department might opt for real-time simulations.

Week 6: Disaster plan auditing and documentation

- Develop the actual look and layout of your disaster plan. You can rely on the three-ring binder if you wish, but many organizations prefer small booklet and brochure formats. Others are working on disaster plan websites and apps. Large libraries can use their intranet to disseminate disaster plan information, but remember that electronic resources might be unavailable during a power failure.

- It's possible that your library has already compiled many sections of its disaster plan. Much invaluable work was completed in preparation for the Y2K meltdown that didn't happen. Following 9/11, many libraries updated or redeveloped their plans. There's no reason to consider that work as a waste when you can revise it for current and future use.

31.6 Characteristics of successful disaster planning committees

- Willingness of each committee member to consider the welfare of your library as a whole before the needs of any particular department.
- Willingness of committee members to work as a team to produce the first version of your library's disaster plan within a specific timeframe.
- Willingness to continue the planning process through regular audits and revisions of the disaster plan.
- Willingness to use plans, procedures, and measures that are already available from internal and external sources.
- Willingness of each committee member to complete specific assignments outside the committee room.
- Willingness of each team member to set an example in taking first aid training, in participating in fire drills, and in helping to distribute disaster plan documentation throughout the library.

Emergency management training for your library: The joys of tabletopping

32

With emergency management training programs that include tabletop exercises, libraries are preparing their employees to survive the worst, and to resume operations quickly.

For openers, threaten your Children's Department with nuclear fallout. Then tell your IT staff that the bomb in their server closet will explode in 1 h. If your senior managers feel neglected, put them in the boardroom with a pot of coffee and tell them to find the best ways to protect your library from floods, fires, and earthquakes.

Who says emergency management (EM) training is dull?

Before 9/11 and the ongoing fears of terrorist activities, EM trainers often struggled to interest audiences in risk mitigation and preparedness, disaster response, and business resumption. Libraries went for years without fire drills. Safety committees met infrequently to discuss ways of grabbing the attention of employees who were more interested in hockey pools and holidays. Even hurricanes and severe winter weather failed to jar people out of their complacency for long.

"We didn't worry so much about emergencies," says Darren, a public librarian in Metro Toronto. "We left it up to city hall to take care of things, and we took an enormous amount for granted. We actually turned down training opportunities because we didn't think they'd be useful."

To be fair, much EM training was slick but superficial. PowerPoint presentations were often vague and soporific, and trainers concentrated on theoretical issues rather than practical concerns. Some trainers relied on motivational speaking to convey notions of organizational preparedness. Others provided mammoth three-ring binders crammed with information, along with workbooks, brochures, pamphlets, and videos. Some library managers felt obliged to review these materials, but their time was limited, and there was so much to learn. Besides, seminars interrupted normal routines and contributed to backlogs.

32.1 On shaky ground?

EM training was often based on unrealistic assumptions and scenarios. For example, libraries on the Prairies busied themselves with orientation sessions that focused on major earthquakes, while a group of Vancouver special librarians spent several weeks making plans to survive the return of the Ice Age. In Calgary, a team of academic librarians pondered the implications of meteorites and crop circles. In Regina, a committee considered the effects of landslides and avalanches on Saskatchewan's urban communities. Of course, a risk assessment should comprise all possible threats and their impact on specific sites and operations. But we must be realistic. An earthquake on the Prairies is as unlikely as an Ice Age in Vancouver, and the Great Regina Landslide is beyond science fiction.

"In some libraries, there was a feeling that we should be more serious about training," says Elsa, a college librarian in Vancouver. "But for a variety of reasons in the 1980s and 1990s, there wasn't a sense of urgency. I've heard that people at a library in the USA held a training session about what staff members should do if they were threatened with alien abduction. Maybe that's not true—nothing more than an urban legend. But it wouldn't surprise me to learn that before 9/11, some librarians were more worried about close encounters of the third kind rather than garden-variety threats like fires and storms."

32.2 Winter weather

Increasingly after 9/11, and in light of disasters such as Hurricane Katrina and the Lac-Megantic train derailment, EM training has become more practical. Trainers have the full attention of their library audiences, which expect comprehensive programs on disasters at specific branches and other sites. Customized presentations are in demand. Trainers know that Toronto libraries must have effective procedures to respond to severe weather, particularly in winter. Methods of dealing with postdisaster transportation problems and emergency communications are essential, along with media management strategies and business resumption plans.

But what's best for Toronto is not necessarily appropriate for audiences in other cities. In Vancouver, earthquake orientation is popular, with an emphasis on protecting library employees and resuming business in postdisaster conditions. On the Prairies, training sessions focus on correct responses to tornadoes, winter storms, and other forms of severe weather. Libraries in Atlantic Canada prefer workshops that cover ways to deal with the full spectrum of natural and technological risks, including power outages and data loss.

Owing to concerns about security, a number of libraries have combined security orientation and disaster response training. Across Canada, library EM trainers have revised their curricula to cover bomb threats, electronic intrusions, and workplace violence. While few librarians anticipate sabotage or terrorist attacks at their sites, none will rule out the possibility of these threats in future. Hence trainers get numerous questions about site evacuation, arson, and handling suspicious envelopes and packages.

32.3 Snow days!

EM training sessions take different forms. Orientation sessions can include a lecture and video presentation covering risks prevailing in a particular community. While libraries offer these sessions to employees from all departments, a growing trend is to provide specific departments with customized training. For example, a school library in British Columbia recently offered staff a seminar on business resumption techniques following winter storms that result in school closures. The trainer reviewed the effects of storms and low temperatures, discussed the subsequent power outage, and outlined ways to communicate with colleagues and students during severe weather. At the conclusion, there were numerous questions about IT problems and the disruption of schedules. Attendees agreed that they were better prepared for winter risks, and urged other departments in their school to take similar training.

"There's no doubt that freak winter storms are more common these days," says Rhonda, an environmental researcher and librarian at a Winnipeg engineering firm. "Low temperatures can continue for lengthy periods, and cause considerable downtime and operational disruption. Librarians have to concentrate on the risks that prevail at their sites, and not be too concerned about what happens in other regions. In a time of climate change, we need to attend to our own specific needs."

Hands-on EM programs are always popular. Sessions introduce employees to tasks such as site evacuation, emergency first aid and triage, working with firefighters and other first responders, and working with civil authorities to inspect a damaged building. Trainers often cover the uses of emergency equipment, and participate in test evacuations. In the past, getting everyone to leave a building could be difficult, since some people didn't take drills and alarms seriously. Now, however, most employees are eager to get involved in any training that enhances their safety. More than a decade after 9/11, the memory of the burning World Trade Centre remains clear in most minds.

32.4 Practical concerns

Librarians worry, however, that hands-on training does not always lead to sustainable preparedness for emergencies. A library can offer a series of springtime sessions on evacuation techniques, but by the end of summer, many employees have forgotten the basics. Some employees have retired or moved to different jobs, while newly hired people can be completely unfamiliar with office emergency measures. Refresher courses can be expensive, and managers might complain about disruptions in their departmental schedules.

The answer in many libraries is the tabletop exercise. The trainer draws up a disaster scenario, and gives it to a group of employees. Often the scenario is kept secret until the exercise takes place, so that participants are forced to deal with the element of surprise and to think on their feet. To make the exercise more challenging, the trainer can set strict time limits. A tabletop exercise can take a day, or an hour, or even a 20-minute coffee break.

Tabletop exercises can have intentionally vague objectives or specific goals. A trainer might ask a group of library staff members what they would do after a fire breaks out in their building and damages a large number of hardcopy and digital items. The group can then discuss damage assessment procedures, "bibliotriage," and the replacement of lost items. Or the group might determine that many items were obsolete and that there was no need to replace them. The results of such exercises can be illuminating as well as surprising.

32.5 Essential questions

An exercise involving specific goals could begin with a brief scenario followed by a request for particular information. In this case, the trainer suggests to a library's Technical Services Department that an ice storm will hit Southern Ontario next December. What is the first message that the department should send to employees? What form will the message take? What message would the department compose for a 1–800 line or a website? Members of the department can have half a day to answer one of these questions, or 20 min to answer all of them. Groups with extensive tabletop experience can work at great speed and produce excellent results.

"Tabletopping" can serve a variety of purposes. It can provide inexpensive EM orientation to a large number of employees in a short time. Some libraries use tabletop exercises for team building; others use them as a way to introduce new employees to the organizational culture and priorities. Exercises can expose weaknesses in a library's disaster plans, and demonstrate better ways of dealing with practical problems.

Tabletop exercises can also reveal opportunities for new efficiencies. For example, a tabletop scenario for a corporate library includes the loss of several stand-alone workstations during a fire. At first, exercise participants are aghast. How will they survive without those workstations, which contained so much vital data! It turns out, however, that almost all of the data in the lost workstations have been duplicated on the corporate network. Some data are obsolete. In fact, none of the lost data are needed for the department to resume business. At the conclusion of the exercise, the library asks the corporate IT group to remove the unnecessary workstations.

"It's easy to understand why librarians like tabletop exercises," says Jim, a retired firefighter and emergency management trainer in Vancouver. "Tabletops are easy to organize. They deliver useful results. They're fast and inexpensive. And they can be developed with a library's specific risks and problems in mind. I don't know of a better all-round training tool than the tabletop exercise."

Libraries that decide to mount tabletopping programs should remember to save all exercise results for future reference. Senior management and auditors can use them to ascertain a specific department's level of emergency preparedness, or an entire library's need for better postdisaster communications with patrons. Here it is essential to remember that nobody "fails" a tabletop exercise. Almost all results are valid and useful in some way, and when libraries use them to enhance employee safety, everybody gets top marks.

32.6 Sample tabletop exercise: A burst pipe causes flooding

Note: The following scenario represents actual conditions.

At 11:30 p.m., a pipe bursts on the main floor of your library branch, and approximately 1400 gallons pour onto the floor and throughout work areas and stacks. Carpets, furniture, collections, and IT hardware are soaked. The flood is not discovered until 6:45 a.m. the next day.

32.6.1 Casualties

Arriving early, one employee slips and falls in a stairwell and sprains her ankle. She lies in the staff entrance for 45 min before she is discovered.

32.6.2 Damage

- Drenched carpets and underlays, which emit an unpleasant odor within 3 h
- Soaked wiring, which increase the risk of fire and computer hardware malfunction
- Soaked ceilings and wall cavities, which emit an unpleasant odor within 12 h
- Wet books—more than half the collection
- Wet digital items—more than half of the CDs and DVDs
- Wet artwork, posters, and ornaments mounted on walls.

32.6.3 Crisis potential

- A reporter from the local media hears about the flood while having coffee at a nearby coffee shop. She arrives in the library and looks for an interview. She calls for a camera crew. The resulting story: "Before leaky condos, there were library floods. There still are. Here's what happened to a branch of our library, and all because of a burst pipe. What are they going to do with those wet computers, those soggy books?"

32.6.4 Tasks for exercise participants

1. Make a list of the first steps that you would take to resume business in your water-damaged work areas. (Take 15 min.)
2. Write a schedule with realistic time estimates for the steps that you have listed in 1 (15 min).
3. Make lists of:
   - the basic operations that you will be able to conduct within 48 h (e.g., communicating with library management at the main branch, alerting patrons, clearing library returns, etc.)
   - the equipment that you will need to offer the basic services listed above
   - official bodies and external organizations that you should contact as soon as possible regarding library service issues.

This part of the exercise will be limited to 45 min.
Please submit the written results of this exercise to the Head Librarian.

Gimme shelter: Extreme weather and your library

33

In his wallet, Craig carries his most prized possession: an old photo of himself as a young man with fists up in a boxing pose. Bare-chested, he wears the standard trunks and boots that boxers wore in the 1960s. He looks confident and strong.

"I won middle-weight fights across the country," he says with pride. "I had a mean left jab that got people talking."

Eventually he ended his career in the ring and started a job with a railway company. He married and had two sons. He became an alcoholic and lost his family and his job, and lives in a tiny room in a run-down neighborhood of Toronto. He was not a regular library patron until, as he puts it, "The weather got really ugly."

Across much of North America, recent winter weather has been colder than usual. One afternoon Craig went for a walk. He thought that he was accustomed to severe wind chill, but as he trudged along his usual route, he felt his fingers and toes go numb. He decided to warm up in a library branch. Pushing open the front door, he saw many people from his neighborhood who had also found the weather hard to bear.

"I saw Rita, a lady who lives in my building," he says. "She was just sitting there, rubbing her hands together to get them warm. She told me that she was scared to stay outside because it was too cold and she might get frostbite, or even freeze to death. Other people felt the same way. It was good to have the library available for us. I sat beside Rita and read a magazine until I warmed up. Then Rita and I made a dash for our building. We didn't go out for a few days after that, and then only to the grocery store and the library."

33.1 Freak weather

In many libraries, extreme weather has forced people such as Craig and Rita to become regular if unintentional library patrons. Scientists and environmentalists believe that climate change has led to freak weather—high winds, unusually heavy precipitation, extended cold snaps, and heatwaves—and that such phenomena will occur more frequently in future. Anecdotal evidence suggests that public and many academic libraries have seen their patrons' visiting patterns change, particularly during periods of extremely hot and cold weather.

"Winters across much of the continent seem more difficult for people to cope with," says Elsa, a college librarian on the Prairies. "We see students virtually camping in our library, arriving as soon as we open and leaving only to attend classes or go home. If there is no space at a table or on a bench, they stretch out on the floor and boot up their laptops. We've seen a number of students with blankets, and even a few sleeping bags."

Years ago, Elsa and her colleagues would not have permitted the use of sleeping gear in the library, but now they condone it—within limits.

"I am not comfortable with students' bringing their bedding to the library," says Elsa. "However, if they're using blankets just to keep warm during very cold weather, I will tolerate it. But if they seem to be setting up light housekeeping in the stacks, I tell them to stop. There's a line that libraries should not allow patrons to cross. This is a matter that senior management should consider."

She notes that peak times in her library are changing. A decade ago, it was easy to predict when large numbers of students would use the library. Mornings would be calm, with growing numbers arriving around the lunch hour. After that, there could be a lull. Around 3 p.m., there would be what library employees called "the rush hour," during which students and faculty members would visit the library to gather materials that they needed for assignments and upcoming lectures. After that, there would be another lull, an increase of arrivals until around 7 p.m., and then a slow decline in numbers until closing time.

"That was the pattern for years," says Elsa. "Now we see unexpected crowds of students in the library at different times. We haven't conducted a survey lately, but I have heard about a number of factors that can influence the behavior of people on our campus."

33.2 On the buses

During the winter, the frequency and reliability of public transit can affect library use. If the buses are on schedule and not crowded, fewer people will show up at Elsa's library. If, however, the buses are slow and crammed with passengers, Elsa will see an influx of students and others in her public areas, consulting their cell phones and laptops, relaxing on any available chairs and benches, or simply lying on the floor. Sudden drops in temperature along with the commencement of snowfall can also increase library visits.

"When the wind whips up a snowstorm, people want to get inside," says Darlene, a reference librarian at a public library in the Maritimes. "Wind chill can be especially nasty in our region, and when people see the lights in our windows, they come in and take a break. We've been serving lots of people who have never been regular patrons. Our director jokes that a bad winter is the best marketing strategy we could hope for. Judging by the number of people who are borrowing items and using the public access terminals for the first time, I'd say that she's right."

New patrons may be welcome in any library, but increasing use of facilities can lead to problems. For example, at a public library on the Prairies, growing numbers of patrons have led to deteriorating sanitation. In particular, washrooms would require more frequent cleaning during and after periods of severe weather.

"I've been a facilities manager for 30 years," says Ed, who looks after the library's buildings. "Our washrooms have been really dirty lately, and the fixtures are showing a lot of wear-and-tear. You don't want to hear the details, but there's no escaping them. We need to check some washrooms three times a day to make sure that the sinks, urinals, and toilets aren't plugged, and that the drains are running properly. Our supplies of towels and toilet paper disappear much faster during cold spells. And the trash receptacles fill up something fierce. We can fill a dumpster with washroom trash in half the time we used to. This situation is really obvious when a lot of people come into the library to get out of the cold."

33.3 Unromantic

Ed must present these facts to the manager in charge of the library's public services. The manager is forced to increase the facilities budget to handle the demand for more washroom maintenance.

"Not very romantic, is it?" says Ed. "You figure that managing a library is all about developing collections and offering sophisticated services and programs. It turns out that some managers have to pay even more attention to a library's physical plant. Cold weather sets in, and lots of people want to use the washrooms. With climate change and extreme weather, this is a fact of life and a professional concern."

While winter has brought colder temperatures, summer months can bring the opposite, in the form of hot spells and heatwaves. In BC, librarians see patrons arrive in their public areas for no other reason than to cool down. Seniors often enjoy hot weather, but it can lead to sunburn, dehydration, and heat stroke. Dave, an Adult Services librarian in Vancouver, believes that seniors require special consideration when temperatures soar.

"I try to ensure that older people have a place to sit down when they arrive at my branch, and I watch for any signs of distress. I've brought people glasses of cold water when they appear to be suffering from the heat, and almost immediately they feel better. I've also called taxis for seniors who don't look as if they'll be able to make it home. It's the least I can do."

Dave notes that hot weather can make some patrons drowsy. While there are rules against patrons' sleeping in public areas, during the summer he tends to leave older sleepers alone, as long as they do not disturb others by snoring. But he has been less patient with people who pass out owing to alcohol or substance abuse. When he saw an intoxicated man sleeping with his head resting on a table, he woke that person up and cautioned him not to repeat that behavior. (Ironically, the man swore that he had fallen asleep because a book that Dave had recommended had been dull.) Dave raises an issue that is on many librarians' minds.

"I wonder how far we can go with the notion that libraries are emergency shelters," he says. "Because of climate change, extreme weather is more frequent, and people might be at risk if they stay outside. But can libraries take on the role of shelters indefinitely, and should they do so in light of the inadequacy of their facilities to handle big, unexpected influxes of patrons? Don't get me wrong. I think that libraries should accept more responsibility when people could die of hypothermia or heat stroke. But are librarians prepared for this responsibility? Do we have effective procedures in place to deal with seniors and homeless persons who might need more than just a place to sit?"

33.4 Convenience

Dave believes that periods of extreme weather will become more frequent in future, and that libraries should work with other institutions in their neighborhoods to assist people who are stranded or otherwise threatened by winter lows and summer heat. He mentions community centers, churches, schools, and sports venues, but admits that all of these have disadvantages. In the end, libraries can be more convenient for many people.

Working in Toronto and Florida, Thomas is an interior designer and a devoted library patron who suggests that, over the next decade, climate change will lead to new forms of library layouts and furnishings.

"There's a lot of work to be done in redesigning libraries to handle changing weather patterns and different kinds of patron use," he says. "As well, we have to take into account the deep budget cuts in numerous jurisdictions, and realize that we can't afford the more costly redesigns and renovations. There's going to be a lot of discussion about what we can do with library space to keep it people-friendly but inexpensive to maintain. There will be controversy."

Thomas holds the controversial belief that empty floor space can be more useful than space furnished with chairs, benches, and tables, especially when large numbers of people arrive in a library to escape severe weather. He is not against furniture *per se*, and notes that disabled and elderly patrons require proper seating.

"But lots of patrons don't need furniture," he says. "Look at the learning commons areas of many colleges and universities. There are lots of students sitting quite comfortably on the floor for lengthy periods. Perhaps they will use knapsacks and coats as cushions, and that's acceptable. The library can give them shelter and access to the usual resources, but avoid spending money on more furniture."

33.5 Great style

Thomas is more conservative about washroom design, and says that libraries should improve maintenance and enhance layouts of the more heavily used facilities. He says that many public library washrooms are too small to handle influxes of patrons. Hence the usefulness of many libraries as shelters is already questionable.

"Good intentions are not enough," he says. "First, librarians and their sponsors must decide how far they will go to accommodate the public when the weather turns bad. Second, they must act on their decisions and plan ways to adapt their space accordingly. And third, they must turn plans into projects. These days, that's easier said than done."

Seated not far from Thomas, a tough-looking man in late middle age reads Hemingway. It is Craig, who has discovered a writer whose books he cannot put down.

"One afternoon I'd had a few drinks and I was tired," he says. "I came in out of the cold—and it was really cold—and I picked up a collection of Hemingway's short stories. I couldn't put them down, and that's saying a lot, because I was at the point where I might have dozed off. Hemingway liked boxing. Did you know that? But I think he was probably a better writer than a boxer. Anyway, he brings me back to this library and takes my mind off everything else. He had a style that got a lot of people talking."

But not even the greatest style will take librarians' minds off extreme weather. It's not something you can return to the shelf.

Section C

Miscellaneous contributions

Bookstore disasters: Surviving the worst

34

You don't need this.

You opened your bookstore this morning and found half your stock drenched from a burst pipe. You can already smell the mold. Large-format items fall apart when you open them. Thousands of pages stick together. Covers warp into wild shapes, making your fiction display resemble a nightmare. You don't need this, but you have no choice. You're faced with one of the more common bookstore disasters.

Insurers and risk managers agree that the best way to deal with any disaster is to be prepared for it before it strikes. Most bookstore floods and fires can be avoided through preventative maintenance. Look around your store. You'll probably find a number of potential trouble spots.

"Booksellers forget that books are among the most fragile consumer goods," says Mike McGee, a fire control specialist and Security Manager at the Vancouver Public Library in Vancouver, BC. "A few ounces of water can render a book unsaleable in minutes. Fire and smoke work even faster. And sometimes fires and water hit you simultaneously. For example, when firefighters open their hoses on a blaze, often there's more damage from water than fire. Considering those kinds of risks, I'd say that preventative maintenance is necessary in every bookstore."

McGee recommends annual inspections of plumbing and wiring, with special attention to any suspicious stains on ceilings and walls. You don't have to be a plumber or electrician to recognize a dripping pipe or frayed wires, but if you discover these risks in or near your store, you should call for skilled help as soon as possible. A plumber's time is expensive, but the bill is not nearly as painful as the sight of warped fiction.

Windows and skylights need regular inspections to ensure that water doesn't seep through cracks or loose panes.

"Skylights can be a real nuisance, especially in wet weather," says David Duthie of Duthie Books in Vancouver. "Skylights are more exposed than other glass, and they tend to leak unless you keep them in good shape. We've had one large skylight repaired once a month during a rainy winter."

If mild Vancouver winters are hard on skylights, how about severe winters in the US Midwest and Northern Europe? A skylight enhances a bookstore's interior, but it also increases the risk of water ingress. If you can see the clouds through your ceiling, be sure that your window to heaven is well-caulked.

Duthie also recommends regular inspections and maintenance of sprinklers and basement storage areas.

"I hate to think about how much stock I'd lose if our sprinklers went off accidentally," he says. "They should be checked once a year, along with the fire extinguishers

How well prepared are you to recover from a fire or flood?

and alarms. Remember that one sprinkler head can soak thousands of books. And storing boxes in the basement should be avoided if at all possible, since water naturally flows toward the lowest point in any building."

Another important preventative measure involves backing up vital computer data regarding inventories, sales, special orders, and other essential operations. Storing the backup disks and other media in a safe location away from your bookstore ensures that you will have access to your data if your computer system is stolen or destroyed. Backing up data is increasingly important for bookstores that might experience power outages and attacks from online vandals and hackers. There is also the risk of human error that results in the accidental erasure of electronic files. It is prudent to keep a full set of current backup data, in case the worst scenario comes true.

Back to your flooded bookstore. You must move quickly. In many cases, the loss of stock is not as inconvenient as damage to walls, floors, and ceilings. Water trapped in wall cavities causes damage to building materials. Presently you might notice an unpleasant odor and brownish stains spreading across a wall. These are signs of significant water damage, and if you can smell the decay, so can your customers.

It's best to hire trained moisture control specialists who are accustomed to restoring retail businesses. Janitorial companies often lack the necessary equipment and expertise to deal with a bad flood. You can't rely on them to get you out of trouble with good intentions and a mop.

"Bookstores are special cases," says Alan Reyno, a moisture control project manager in Western Canada who has worked on dozens of damp and fire-damaged retail sites. "If I can set up dehumidification units on site a couple of hours after a pipe bursts or the roof leaks, then I can decrease building restoration costs substantially. As for books, my success will depend on the quality of the paper, the amount of time the book has been damp, and the way it has been handled in its damaged condition."

Reyno and his colleagues in the profession have dehumidified and freeze-dried books and other paper items with excellent results. Unfortunately, the restoration of paperback stock is not cost-effective, and many soggy thrillers must be either discarded or sold at greatly reduced prices. Nevertheless, reduced revenue is better than none, and public interest in your fire sale or post-flood customer bonanza could increase overall sales.

Moreover, there are ways to profit from adversity. Years ago, a clever Toronto bookseller sold a large collection of smoke- and water-damaged stock to an American university that needed soggy books for its paper conservation laboratory. Thus a bookseller's disaster contributed to higher education, while the bookseller recovered his losses in full.

Insurance, however, is the usual source of postdisaster funds for repairs, and most North American booksellers have policies that cover different risks. Be honest: have you read your store's insurance policy? Can you describe exactly what risks it covers? If you're not sure how your policy protects—or fails to protect—your store, read it carefully and call your broker if you have any questions. Remember that an insurance policy is a legal contract with certain limitations, and the insurance company will not give you a check to repair damage that the policy does not mention.

You may have lost much of your stock, and your store is in need of serious repairs, but the manager of moisture control team tells you that he has seen much worse problems, and the soggiest or most burned-out bookseller can usually find some way to stay in business. He might also tell you to develop a preventative maintenance program for your store, and never to ignore an old pipe that drips.

You've learned your lesson. Let's hope that you won't need another.

A word from a pro: Protecting your store against book thieves 35

As a professional book thief, I've become well-acquainted with your store. I know the layout and blind spots. I've scrutinized the rear exit and your security camera system. But what interests me most are the large-format hardcovers on display by the front door—fine arts items priced at over $100 each. This is the season for high-quality merchandise. Mind you, books of all kinds are constantly in season for a book thief.

Who am I? The cops will tell you that I come from numerous backgrounds. I'm a little boy in a bomber jacket and an elderly woman in a shabby raincoat. I'm a promising undergraduate at a local university; sometimes I'm a senior professor. I'm a street person, a housewife, and the vice-president of a multinational firm. I'm a politician, a priest, an addict, a lawyer. I'm your best customer, an occasional browser, and somebody you've never seen before.

I have different motivations. I might need the money that I can get from reselling your stolen stock: cash for heroin, food, rent, Christmas, debts. Or I might need the books I steal for educational purposes. As an undergraduate science student, I'm angry about the high prices I'm expected to pay for my chemistry and physics texts. I need them to pass courses, graduate, and become a productive member of society. By charging such high prices, you're standing in the way of social progress, so I'm justified in ripping you off. At least that's what I believe as I walk out of your door with those texts under my coat.

Another motivation for book theft that booksellers will understand as quickly as they deplore it is intellectual compulsion. I want that book by Kant or Hegel or Marshall McLuhan because it's worth studying. I can probably afford it, but I have so many other expenses—MasterCard, the phone bill, my dry cleaning… and I want that book now. It's so easy to drop it in my shopping bag.

No matter who I am, no matter what my motivation, I'm a thief and I'm about to rob you. How? Pay attention.

Successful theft depends on three steps. First, I must target the items that I want to steal: what they are and where they are located in your store. Second, I must conceal those items so that neither you nor a customer will suspect me of theft. Third, I must exit the store without being stopped, questioned, assaulted, or arrested. Once I'm outside and a short distance from the store, I'm safe. It's difficult to prove anything against me when I'm down the street; moreover the cops have enough to occupy them without having to work up a case against some old guy who allegedly shoplifted a copy of *The Medium Is the Massage*. What the police might be interested to learn about the old guy is that his apartment is crammed with one of the better private collections of media history, all of it stolen.

Robertson on Library Security and Disaster Planning

I have various ways of concealing items that I intend to steal. If the item is small-format and thin, I can pop it under my jacket or into the pocket of my overcoat. I can drop it into my shopping bag. Or I can slide it under my arm and, assuming a purposeful air, walk out through the door. Do I look like a thief to you? Surely you're joking.

"In-store ownership" is a popular shoplifting technique among students. While you're not looking, I remove the price tag from the book I want, then scribble my name on the flyleaf. Time permitting, I might use a highlighter on a page or two. Perhaps I'll artificially age the book by cracking its spine and giving the cover dog-ears. By the time I make my exit, the book appears well-used. Was I supposed to leave it at the counter when I entered your store? Sorry, I didn't see the sign. I'm distracted by exams. I've lost my glasses. I'm in love. Anyway, this is obviously my book, as you can see from all the personal touches. So long, sucker.

I can work with a partner who will divert your attention while I walk out with a pile of your best stock. I can switch price tags: does it surprise you that this recently published folio atlas is on sale for $2.99? I can sneak out through your back door; if you catch me, I'll say that I was looking for a washroom. And why am I carrying the latest Booker winner? Because I'd like to skim the first chapter sitting down. Must you ask so many personal questions?

Remember that a professional thief is adept at feigning innocence and redirecting your attention. You can demand that the old lady open her coat and shopping bag: she'll accuse you of sexual harassment. You can tell the kid with the skateboard to stop switching price tags: he'll swear that he was merely replacing a tag that fell off. If you physically restrain a street person before he can exit with a book concealed under his coat, he will accuse you of assault. And there's always the risk that the thief you attempt to restrain is carrying a knife or other weapon. No wonder booksellers often make little effort to prevent theft in their stores.

I don't want this to get around, but there are effective ways to stop thieves like me. You may already know about mechanical measures such as security cameras, mirrors, tattle-taping, and door sensors, all of which discourage amateurs. But a professional thief knows that you haven't the patience to keep your eyes on the camera monitor and mirrors continuously. Professionals know that you don't Tattle-Tape your paperback stock, and that a thief detected by a door sensor need only show that he's carrying a tattle-taped item from a public library to be permitted to exit.

To frustrate a professional thief, you should try to give every customer in your store the impression that you're constantly present and prepared to answer questions and give advice. I don't like bookstores that offer high levels of friendly, attentive service. I don't know who's watching me, and in a store with an open layout featuring fewer nooks and crannies, I'll have a problem concealing the books I want to steal.

I avoid stores that have a counter beside the front door, angled in such a way that the counter clerk can see most of the shelf area in a single sweeping glance. That clerk can also get a good look at me as I enter or exit, which makes me uncomfortable. So do clerks who wander around the shelf area and offer assistance. So do stores with securely locked back doors, good lighting, and locked display cabinets for the most expensive stock. Note that sign: "Please ask the staff if you'd like to examine these rare editions." Give me a break!

Another tactic that annoys me is the placement of comfortable chairs in the shelf area. Very clever: you enhance the cozy atmosphere of your store while turning honest customers into unwitting guardians of your stock. For you can be sure that I'm as uneasy near some character perusing Atwood in an armchair as I am of a cheerful circulating clerk. They interfere with concealment, they get in the way, and I can't trust them. Stores with a seated clientele are simply not worth the trouble.

No matter what kind of security system you implement, you will never stop me altogether. As a professional, I'm always finding new stores to hit. But those of you who get smart can slow me down, reduce my take, and eventually drive me away. That's not very neighborly, but you won't miss my patronage.

Sabotage for beginners: Destroying vital records

36

Insurers can be easy victims for saboteurs who recognize the industry's dependence on information resources.

Looking for a way to wipe out a big insurer? Or bankrupt a broker? Perhaps you'd prefer to wreak havoc in the office of an adjuster. One method is guaranteed to attain your objective. Simply destroy your victim's vital records. Without them, a company is in serious trouble, and sometimes out of business.

Remember that the records you destroy must be vital. Companies produce tons of paper that is worthless after its first use, but their vital records are necessary for ongoing business purposes. Don't bother with the duplicates of memos regarding last year's golf tournament, the multiple copies of the strategic plan for 1972, and the documentation for the mainframe that was replaced decades ago. Concentrate on the information resources without which your victim's operations will come to an abrupt halt. Current policy files, claims files, human resources data, and payroll figures are some of the more valuable records. Lose a small quantity of these, and a crisis will be difficult to avoid.

After all, what can a customer service representative say to a policyholder who phones for an update? "I'm sorry, sir, but we've lost your file. That's right. You've been paying premiums for years, and we have no idea what for. Maybe we have your address, maybe not. But at least we have a phone book. What was your name again? Hello? Hello?"

It's unthinkable. But vital records go missing every day, in offices across the continent. And hapless managers find themselves trying to explain the indefensible to the CEO or senior auditor: "I don't know why those claims files ended up in the shredder. I didn't put them there myself. I'm very sorry about the angry calls and threats that our receptionist has been getting… Backup copies? Let me wring out my shirt, and I'll think of a reason that we have no current backups."

Whereas 25 years ago, your victim's vital records were mostly paper files, nowadays they can be digital, on microfilm and CDs, or in a combination of media. File rooms have become records and information centers, and mountains of document boxes have moved from head office basements into high-security warehouses. But the increasing complexity of corporate records systems does not necessarily lead to smarter or more sophisticated records users. In fact, the key to successful sabotage can be a records system that is too complicated to be managed securely.

But first, before developing your plan of attack, consider how vital records deteriorate without sabotage. Modern offices use enormous quantities of less expensive paper, because only a small percentage will be retained for longer than a few years. Even the best paper will wear out, especially with regular handling, and inks fade when

Robertson on Library Security and Disaster Planning

exposed to light. Computer printouts are notoriously unstable, with the ink fading as quickly as the paper yellows. Printouts are often illegible within 5 years.

Digital records are even more fragile. Disks and CDs are very sensitive to heat, and will corrupt after even a brief exposure to high temperatures. Smoke and dust particles can damage any electronic equipment and media. In the absence of these risks, you can always rely on rough handling and inadequate storage to cause the desired trouble.

In targeting digital records, the saboteur never works alone. Reports of hackers, viruses, worms, Trojan horses, and time bombs are increasingly frequent. It seems as if corporate computers are under constant attack. If you don't wipe out an insurer's vital data, rest assured that somebody else will attempt to do so shortly.

You can also depend on human beings to destroy records unintentionally. They trip over a cable in the systems operations room, and disable the corporate network. In demonstrating the skill with which the Arsenal striker passes the ball, an underwriter knocks a hard drive off a desk and fumbles a database. Meanwhile, highly intelligent and usually reliable systems staff mislabel tapes, lose vital documentation, and delete the wrong digital files. The VP of Information Services will reprimand them when he returns from his annual conference. But he cannot be overly harsh, since he left his laptop in the London airport lounge, and is embarrassed to admit that he has lost his company's new marketing plan.

As any insurer will tell you, human beings are often disinclined to mitigate the natural phenomena that pose constant risks to information systems and vital records. Consequently, fire and water destroy countless paper records. Drenched paper decays quickly unless dried out before mold forms. And most often the source of the water is not a firefighter's hose. Roof leaks, cracked skylights, windows accidentally left open, faulty plumbing, and malfunctioning sprinklers are the most common culprits.

With so many advantages, the saboteur can cause major losses of vital records by the simplest means. In fact, successful saboteurs will try to make such losses appear accidental, or the result of purely natural forces beyond the control of office staff. Over the weekend, burst pipes cause a flood in the Information Services operations room. Four servers have been soaked, and are probably irreparable. Unfortunately the backup tapes are missing. The company has lost all customer data input during the past quarter. And only the saboteur need know that he was the one who cracked open the washroom pipes and stole the backup tapes, which he found on a shelf in the operations room.

The saboteur can activate the sprinkler system to achieve similar results. Or he can disconnect the uninterruptable power supply (UPS) and then knock out the power to the building. What caused that power failure? It can remain a mystery as long as the saboteur takes basic precautions and flicks the right switch. The results can be thrilling: large amounts of data have vanished. How much? Because few companies keep up-to-date data inventories, it's hard to say. But the saboteur can be certain that much of what was lost was probably vital.

The hacker is to be commended for many of his techniques, but note that he requires an in depth knowledge of computer systems and the Internet. Moreover, to get around a company's electronic firewalls, he will need knowledge and patience. The old-fashioned saboteur needn't know how to turn on a PC as long as he's prepared to take a heavy wrench to a washroom pipe, or to pull a few plugs from UPS sockets, or to leave a window open during a rainy evening.

A loss of vital records leads to nasty results. Auditors and government agencies ask numerous questions. Senior managers become ornery. People can be demoted, sometimes fired. Others must work hundreds of extra hours to replace or reconstruct lost data and paper files. The saboteur will be relieved to learn that the police are rarely consulted. They would ask too many embarrassing questions, such as "Why don't you store up-to-date backups in a secure off-site location?", and "Can't you tell which data are missing?", and "Don't you lock your file cabinets at night?" Besides, the police are busy chasing villains who steal laptops full of confidential information from unlocked cars. It's a growing crime industry.

Some companies pose more of a challenge to the saboteur; others completely thwart him. The company that succeeds in avoiding most vital records losses has an efficient records management system that emphasizes security and efficiency. A top-notch records manager will be aware of all potential risks to a company's vital records, including sabotage, human error, and natural disasters. While it is impossible to anticipate all contingencies, the records manager can protect vital records by identifying them, noting their locations, and backing them up so that they can be recovered in their entirety following any kind of loss.

Saboteurs should note the signs of a company that will not be easily victimized. First, employees take seriously the corporate security policy. They clear their desktops at night, lock up their file cabinets, and back up the day's more important data. As much as possible, employees avoid traveling with sensitive paper and data. The corporate website does not provide easy access to vital records without a comprehensive security routine. Second, the records manager has identified the company's vital records and can spot and counteract any deterioration or loss quickly. Third, the records manager is always looking for better ways to protect and back up information stored on vulnerable media such as aging paper. While insurers still depend on paper for many purposes, the records manager is eager to convert it to a digital form that is easier to back up. Many companies still rely on microfilm, which, while is not as easily backed up as electronic data, is even more physically stable.

The saboteur is prudent to find a more willing victim than a company whose employees are security-conscious and whose data are effectively backed up. It is a waste of time to destroy information resources that can be replaced with little cost or difficulty. You might as well find a softer target. Fortunately, there are more than you can count.

36.1 Saboteur's bane: High-quality off-site storage

Top-notch off-site storage facilities can be serious impediments to the saboteur. Those intent on destroying an insurer's backup media held off-site could be thwarted by a storage facility with the following features:

- *Flexible media accommodation:* The facility can receive, store, and conserve the volume and kinds of media that the insurer uses for backups.
- *Acceptable design standards:* The facility meets design standards that have been recommended by trade and professional associations such as ARMA International. The facility conforms to all local building, security, and fire codes. In seismic zones such as California and British Columbia, the structure of the facility can withstand earthquake loading.

- *Secure transportation:* Backup media are transported to and from the facility in unmarked vehicles. Drivers have been instructed not to allow unauthorized personnel to ride in the vehicle or inspect its contents. Drivers will not answer questions about their schedules or destinations.
- *Trained personnel:* All facility personnel are trained in the handling of backup media.
- *Security measures:* All facility personnel have passed a security check. The facility has access controls including alarms and electronic monitoring. Access is restricted to facility personnel; insurers can inspect their media in a visitors' area.
- *Access provision:* The facility can provide 24-h access to backup media in case of emergency.
- *Strategic alliance:* The facility plays an essential role in the insurer's security and disaster recovery plans.

Saboteurs should remember that significant disruption of an insurer's operations depends on the destruction of both the original electronic media and their backups. If backups are allowed to survive, the saboteur will ultimately fail.

Success without college: How cheaters prosper using fraudulent credentials

37

He looked good. He sounded convincing. He impressed the CEO, and earned promotion after promotion at one of the larger insurers. Colleagues speculated that he'd be a VP soon. Everyone agreed that he had the right qualifications, including a Harvard MBA and a PhD in Economics from the University of Chicago: noteworthy achievements for a man barely 30 years old.

Enter the spoiler: a woman recently hired to work as an executive assistant in the insurer's Human Resources Department. Hardly an academic star or a climber, nevertheless she had sound intuition and first-class research skills.

The HR manager asked her to collect biographical data for the company's annual report, which was to feature profiles of the insurer's most talented managers. She consulted the files and found the usual documents concerning the young super-achiever's education: photocopies of degrees, letters of reference, and Certificates of Excellence from various professional associations. Only a photo was needed to complete the package. Did the young man have a good shot of himself, perhaps standing in Harvard Yard in his mortarboard and gown? The executive assistant called his office, but he was on holiday. So she called Harvard and the University of Chicago to see if they had photos in their alumni files. No one had ever heard of the brilliant young man. There was no record of his attendance at either university.

Surely there was some mistake. After all, he was reputed to be a genius. But the assistant was suspicious. She decided to investigate further. What she discovered was a potentially enormous embarrassment to the insurer and its CEO. The young man was a fraud. Clever, personable, eager, and creative, but in many ways less qualified to fill a senior position in the company than the woman who caught him.

Certainly he had worked in the insurance industry for years, and knew about different products and marketing. He kept up with management trends and was effective in dealing with brokers. But he had lied and cheated to further his career, and used bogus tools to enhance his image.

First, those degrees. Harvard is a world-class institution, and the University of Chicago boasts a regiment of Nobel Laureates on its faculty. Anyone with credentials from these universities is likely to make more of an impression than some honest soul with a decent diploma from the University of Northern British Columbia. The latter institution may be respectable, but too new to be revered as a training ground for the elite. It is easy, however, to make a counterfeit copy of a degree from any college or university. All a fraud needs is a scanner and a good graphics program on a PC. Using

How cheaters prosper using fraudulent credentials

these, a fraud can produce almost exact copies of any degree, suitable for hanging on the office wall, or in the executive suite. Rarely do HR departments check out a job applicant's degrees to confirm that they're real. Often a photocopy of the degree is all that is needed to satisfy interviewers.

Frauds use scanning technology to produce different kinds of bogus documents, including university transcripts (with straight As of course, and perhaps a "Summa Cum Laude" designation that sounds so classy), certificates of merit and achievement, awards, and letters of reference.

How glowing letters from the three most respected insurance CEOs in New York? These could be more valuable than the best degree. The fraud's first step is to obtain the appropriate corporate stationery, which insurers distribute by the ton. Lacking unused stationery, the fraud can work wonders with a color photocopier. A visit to the business division of the local public library will turn up the right annual reports, which contain the CEOs' printed signatures. Next, the fraud must concoct glowing remarks about himself to include in the letters: "Rarely in my 30 years in insurance management have I met anyone with more promise, enterprise, and integrity. Mr. X can guarantee success through sheer force of character," etc. (In fact, force of character might be Mr. X's only real qualification.) Finally, the fraud combines the ingredients with credible dates, and sends them to prospective employers.

Some frauds simply modify or exaggerate the credentials that they already hold. An MBA from the University of Calgary with a concentration in Finance is transformed on his resume into an MBA with a concentration in insurance marketing and administration or a mail-order Commerce degree from Southern Pacific Christian University (which offers no courses and asks no questions) becomes a Master's in Economics with an emphasis in demographic trends. Frauds often take for granted that no one will check their personal data and educational records. They will even claim authorship of academic theses, dissertations, and other publications that are fictitious, and they can be confident that nobody will try to find copies of their works. After all, who wants to read a boring thesis? As long as the title looks authentic on a job applicant's list of publications, most HR managers will accept it as real.

The reference check should be the moment of truth, virtually impossible to fake, but recently in Toronto a fraud demonstrated how to deal with calls to persons listed as his references. Before the HR manager had a chance to phone a Claims VP to find out more about a particular applicant, that VP phoned her, or so she believed.

"I'm going out of town soon," he said, "so I thought it best to call and tell you how impressed I was with that fellow, who has applied for a job with your company. He's a superb candidate, one of the top men in the insurance business. I'd be happy to recommend him to any outfit looking for a bigger share of the market."

Weeks later, the applicant who had received such high praise admitted under pressure that his father-in-law had made the call and pretended to be the VP. The father-in-law was a retired carpenter with chutzpah and a powerful baritone voice. Had the applicant's bitter ex-girlfriend not exposed him and revealed his methods, he might have become a senior manager on the strength of old-fashioned telefraud.

A more recent trend in credential fraud is the personal website loaded with self-aggrandizing half-truths and outright fabrications. Often attractively designed, such websites can contain misleading photographs. There might be a shot of the fraud standing outside the front door of a large insurer's head office. The caption reads: "My home away from home." He has never worked for that insurer, but the shot is impressive. And there's no law against standing outside an office block. Another shot shows the fraud standing at a podium, apparently delivering a speech. Caption: "Another keynote address to a large audience of brokers. They liked my message." Of course the viewer does not see the audience, because there was no audience. The shot is as phoney as its subject.

Along with photos, fraudulent websites also include fake testimonials in bold type. "Mr. X is a champion. With men like him at the helm, the insurance industry will remain strong." Who said this? Perhaps a retired prime minister, or a former minister of finance, or Conrad Black, whose name appears under the quotation. But neither the politicians nor Mr. Black said this. They have never met Mr. X, and don't know that he exists. Mr. X doesn't mind, as long as the HR manager viewing the website is convinced that he has high-powered connections.

What happens to frauds if they're caught? Rarely, if ever, do companies prosecute job applicants or employees who have presented false credentials. Sometimes the matter is ignored; in other cases, the fraud is asked to leave as quietly as possible. Occasionally a buyout is arranged, or an early retirement, or a transfer to another office and a different position. Above all, companies strive to avoid the attention of the media. Some buyouts involve a promise of silence regarding all aspects of the case. Unfortunately, the disinclination to expose frauds in public allows them to repeat their activities elsewhere. Some frauds have worked across the continent, always walking away from employers who catch them.

As for that brilliant young man who wasn't from Harvard and the University of Chicago, he used his buyout money for an extended holiday in Grand Cayman. Rumor has it that he returned to Canada early this year, but nobody knows what he's up to at present. By the way, has your company recently employed a young Ivy Leaguer? Or somebody who talks about the Nobel Laureates he met when he lived in Illinois? Watch that man. And check him out.

If the credentials are too good to be true, they probably aren't.

Human Resource managers are advised to watch out for the following:

- Job applicants with too many qualifications. Very few people have two PhDs from different world-class universities.
- Job applicants who are too young for their credentials. How many MBA holders are 21 years old?
- Overnight sensations, such as employees who have never attended university but who return from holidays with freshly printed Master's degrees for their walls.
- Personal chronologies that don't make sense. For example, it is virtually impossible to work full-time as a Chief Financial Officer in Winnipeg while completing degrees at universities in Edmonton and Toronto, and raising a family in Halifax. But a fraud might tell you how much he enjoys the commute.
- Job applicants with too much experience at companies that have gone out of business, or whose officers are deceased. Dead references should automatically raise warning flags.
- Famous persons whom you've never heard of. A fraud might claim that he has worked with a series of prime ministers on various national policies, and mention that he's next in line for an Order of Canada. If he's this good, somebody in your company will have heard of him. If not, beware.

38 Social engineers and their victims: Making the wrong contact

You'll arrive in this airport lounge any moment. Between flights, hundreds of senior managers and other important people stop in for a drink. I'll know about you from the tag on your briefcase, which contains your business card. Even before I approach you, I'll know your name, title, and company. I'm delighted to learn that you work for a big insurer. We must talk.

I'm what security experts call a "social engineer." Using devious methods, I obtain confidential information from unwitting sources. I can sell it to your competitors and anyone else who's prepared to pay for it. While I concentrate on covert information gathering, I might also get involved in other criminal activities including investment scams and fencing stolen goods.

But don't assume that all social engineering is illegal. Idle curiosity and the love of gossip turn many of us into social engineers. To a professional like me, however, prying corporate secrets out of unsuspecting strangers can be profitable. It's safe: you rarely see my kind in court, unless we're charged with other offences such as fraud. In fact many of my victims are too embarrassed to admit that they've been "engineered." Nobody enjoys looking like a sucker.

Why are you my target? Because someone in your high-powered position probably has access to much valuable information. You can tell me about the "six Ps" of your company: plans, projects, products, prices, purchases, and personnel. Your competitors would love to know the marketing strategy for your company's new products, and when you plan to release that new homeowner policy. What's the premium? And what about the new software packages that you're developing to manage your customer service department? Who will you hire to run your projects, manage your databases, and handle that big merger?

Think of it this way: you'd be delighted to have this information about your competition, wouldn't you? Now you understand why your competition would like to know about you, and why certain persons who work for your competitors would be happy to pay a social engineer to dig up useful intelligence about your own company's six Ps. You also realize why social engineers engage in unethical behavior.

I'll look for you in different public places. Bars and restaurants are hotspots for social engineering, as are convention halls and trade fairs. Airports, train stations, and bus depots are traditional favorites, partly because travelers must display their identification freely to enter departure areas. (That's a fine passport photo, by the way. It really captures your character.) Any lobby, waiting room, or Starbucks could be a base for my operations. These are natural places for people to meet and talk, exchange views, gossip, and information concerning your company's buyout of an American insurer with substantial Asian assets. Hi there!

Robertson on Library Security and Disaster Planning

Successful engineering starts with an effective opening line. Simplicity and fluency are essential. Ideally I'll manipulate you into an extended conversation that begins innocently and moves into confidential matters before you realize it. Let's say that I see you checking for messages on your BlackBerry. Here's my opening:

"Do you mind if I ask you if you like using that thing? It's like a handheld PC, isn't it?"

Yes, it is.

"What can you store on a device like that? I mean, can you carry statistics and that kind of thing?"

Yes, you can.

"But isn't it hard to read that stuff on such a tiny screen?"

No, not as hard as you think. Look at these data concerning my company's new products.

"Hey, you can see the projected sales figures really clearly."

You can. And just look at the way that we've organized the figures for New York.

"Look at that! That's amazing. You got time for a cup of coffee? I'd like to check out that machine a bit more. There's a Starbucks around the corner."

And for a cup of coffee, you find yourself sharing sensitive information with a stranger about whom you know nothing. You're won over by his enthusiasm and apparent naiveté, which inspires you to show him information that you might not share with even your closest associates. Does this sound farfetched? In fact it's an approach that a social engineer used recently at airports in Vancouver and Calgary, and at a conference in Banff. It's hardly original. In the world of social engineering, it's equivalent to "Do you come here often?" in a nightclub.

Engineering victims are often highly intelligent people who are amazed that somebody they met in an airport persuaded them to reveal confidential information. How could they trust somebody they barely know? While we might not want to admit it, most of us are too easily disposed to trust strangers, especially when they tell us intriguing stories about themselves. They can be convincing and dramatic. Often they open their bogus autobiography with statements that pique your curiosity. For example:

"I've had a hard life."

I'm compassionate. Tell me more.

"After I left Harvard, I didn't know what to do with myself."

My company employs several Harvard grads in management positions. A person with your education and background must be trustworthy.

"My family was rich, but money isn't everything."

An idealist! You'd never stoop to dirty practices, would you?

"I represent a number of offshore interests."

A well-traveled Ivy League idealist who might know several of the senior managers in my company. Here's somebody who deserves trust.

"They gave me a medal for what I did in Iraq, but that's another story that I don't want to talk about now. It's too painful."

I'm sorry. Perhaps we should stick to the sensitive report on claims management that nobody except my CEO has read. Then you could tell me about your adventures in the Middle East. A hero, too!

Meanwhile, the liquor flows. It loosens your tongue and makes you say things that you shouldn't. It also forces you to answer nature's call. When you leave the table to visit the washroom, you might neglect to take along your briefcase, laptop, or BlackBerry. But that Harvard grad with the Iraq medal seems so dependable. You don't start to worry until you return to find that your drinking partner has disappeared with everything you left behind. This includes the reports that were for the eyes of your senior managers only.

Even more embarrassing is the admission that you indulged in illicit substances with the stranger who stole your information. The Harvard hero offers you marijuana or cocaine. If you accept, you're taking a big risk. Dope can be much stronger than you expect, and it can lower your defenses even faster than scotch. You might hand over confidential files without even realizing what you're doing. Worse, you could wake up in a strange place without your wallet, passport, and briefcase. It happens frequently. To avoid bad press, companies avoid making a police report. You understand why social engineers get away with so much.

Another engineering tool is the phoney celebrity contact. You're great company, especially when you reveal the dates of your upcoming product releases. Perhaps you'd like to join Mick Jagger and me for a drink next week? He's coming to town with Lady Gaga to do a charity concert. You haven't heard about it? Well, it's pretty exclusive. Only the right people are invited. Say, those new products: what kind of premium structure are you looking at? It's confidential, but I'm curious. We won't have a chance to talk about it when the music starts. I'll make sure that you get a ticket.

Sometimes the engineer claims to be a celebrity himself. He might claim to be the author of a book that's about to be published. It will reveal the truth about Wall Street. Or Microsoft. Or the pharmaceutical industry. Would you like a signed copy, and to join the special group of insiders who find out about these things before anybody else? Social engineers are adept at making you feel privileged and exclusive. You might want to return the favor by revealing something that only the senior management of your company should know. To make themselves even more interesting, engineers tempt you with attractive offers that you'd never expect.

How can you protect yourself and your company? Staff orientation is the most effective way to ward off social engineering attempts. Make all employees aware of the risk, and inform them about the engineer's techniques. Remember that any employee could be a target, from the CEO to the young person in the mailroom who opens and distributes sensitive letters, proposals, and reports. Your receptionist can be especially vulnerable, since he or she knows who's in your building and has a good idea of why they're there. The receptionist might also be in charge of a corporate fax machine and every piece of correspondence that arrives in its tray.

Forewarned is forearmed, but employees should also be encouraged to tell their managers about any attempt at social engineering. Any employee can attract an engineer, so it's wise to be vigilant.

Because I'm still waiting for you, and we're bound to cross paths.

State of the art: Risks to valuable artworks and their owners

39

Owning and dealing in art involves risks including fire, flood, and fraud. What are the best ways to protect treasures and investments?

Sadly, most art disappears over time. War takes its toll, as do fires and floods. Thieves remove inadequately protected masterpieces from galleries, offices, and private residences. Owing to the weaknesses in their creators' materials, some paintings and sculptures simply disintegrate. It is surprising that many valuable and culturally significant artworks survive for as long as they do.

Artists and gallery curators are painfully aware of the leading cause of losses.

"I'd say that neglect is the risk that exacerbates all other risks," says David Goldie, a Canadian landscape painter whose canvasses hang in corporate offices across North America. "Neglect is essentially human error, and no matter how much one person values a painting, there are countless others who consider it mere decoration, and don't care about it at all."

39.1 Poor storage

Examples of neglect include hanging fragile pieces in direct sunlight that can fade watercolors and drawings in a few months or less. In a Toronto boardroom, a series of charcoal portraits discolor quickly owing to the warm, dusty air from nearby heating vents. Meanwhile in basements, attics, and storage cupboards, artworks warp and become moldy.

What's at the bottom of that pile of old phone books? Perhaps a Dali print that a senior manager bought at an auction in the 1950s, or an A.Y. Jackson drawing that a former CEO received as a gift from a business partner. It's a shame that the surface of such a treasure has been so badly abraded that it's not worth restoring.

"Poor storage is common even in large and famous galleries," says Goldie. "Collections grow, and space becomes scarce. Curators tell work crews to store items temporarily in a poor quality space, and what was supposed to be short-term becomes long-term storage in a corner that eventually gets damp and attracts insects."

39.2 Inherent vice

Poor environmental conditions can act as a catalyst for an artwork's inherent vice, which is the physical weakness of its materials that will cause it to deteriorate. A notorious example of art damaged by its inherent vice is the entire oeuvre of American artist

Robertson on Library Security and Disaster Planning

Albert Pinkham Ryder (1847–1917), whose paintings are falling apart owing to his unwise practice of adding alcohol, candle wax, and bitumen to his oils. He also poured copious amounts of varnish over everything, and mixed slow-drying and rapid-drying paints. In his old age, Ryder's personal habits declined. His New York apartment filled with trash, and dust coated every surface, including those of his paintings, many of which started to show signs of their inherent vice shortly after Ryder's death.

Conservators can recommend effective ways to protect fragile artwork, even those with advanced inherent vice. They are powerless, however, when neglect and other circumstances result in a fire or flood. In many modern buildings, a fire is more likely to result in water damage than charring. Sprinkler systems can douse flames and everything else in the vicinity with hundreds of gallons of water in seconds. The fire may be out, but water shot under high pressure from a sprinkler can ruin even the sturdiest painting. If the paint remains intact, the surface that it coats might sustain water damage. It takes no more than a few hours for water to encourage mold growth and harmful chemical changes. No one doubts the usefulness of sprinklers, although there are art collectors and gallery curators who wish that they could rely on fire controls that didn't pose such a risk to their collections.

39.3 Managing gallery risks

In light of the increasing variety of prevailing risks, some gallery directors are demanding higher standards of risk management for their facilities. Jan Ballard, executive director of Buschlen Mowatt Galleries (BMG) in Vancouver and Palm Desert, California, insists on the latest security techniques and equipment for her public viewing and storage areas.

"We display museum-caliber art, and don't compromise when it comes to security," she says. "Our front door is solid steel, and our alarm systems are first rate. We've been in business for 30 years. We've experienced only a couple of very minor security breaches, and we've never sustained a serious loss."

BMG's display areas are open and well illuminated to enhance the viewing of paintings and other artworks. Visitors might not notice the physical security system, but it's constantly in operation. There are always staff members present to answer questions and discuss different pieces. They're also on the alert for anything unusual. The layout of the display areas allows staff members to keep every corner under surveillance. Entrances to storage areas are locked and inconspicuous. But while BMG is more secure than many bank branches, its display areas don't feel like a vault.

39.4 Smash-and-grabs

The most common security problems for North American galleries include smash-and-grab thefts and burglaries. Some thieves act like shoplifters, and try to leave galleries with art under their coats. Gallery directors suspect that occasionally collectors employ criminals to steal particular works. There have been attempts to bribe gallery staff members: "Leave that painting by the back door at closing time, and I'll give you $200."

On several occasions, thieves have triggered a gallery's fire alarm and stolen paintings while the gallery staff evacuated by the rear door. But chances are that none of these techniques would be effective at BMG, which has planned for different kinds of theft attempts.

Ballard notes that proper preservation of art begins with an awareness of the specific environmental threats to her gallery. Her staff members recognize the moisture ingress and humidity problems that many modern buildings face, and look for any evidence of water in their workspace. The office building in which BMG's Vancouver gallery is located is well maintained, with regular inspections of the roof, windows, and plumbing.

"Our building managers know what we need, and we haven't had any flooding," says Ballard. "Of course there's always the possibility of a leak, but we take the necessary precautions before there's trouble. Ongoing risk management is part of our standard operating procedure."

39.5 Professional management

While some galleries have a reputation for inefficient business practices, BMG is professionally managed. A controller and accountants oversee accounts, and the gallery is run as a business rather than an art lover's dream or an artist's sideline. Ballard emphasizes the importance of proper inventory management. She mentions BMG's painstaking handling of artworks. From the moment a painting or sculpture arrives at the gallery's loading bay, specially trained staff members treat it with great care. Unpacking an expensive painting involves a more demanding series of procedures than removing a new PC from its box. Moving a large painting from its crate to a space on a gallery wall can take hours. Frames and display cases must be clean and properly constructed.

"The management of a gallery involves many small steps," says Ballard. "Leave out a step and there could be problems. But I believe that BMG would do well in any insurance audit. We insist on good business practices, as well as the most stringent environmental and handling standards."

39.6 Fraudster in action

Dealing with a professionally managed gallery is one way to avoid fraud. Over the past decade, a number of North American companies—including several insurers—have fallen victim to a bogus art scam that has cost substantial amounts of money and embarrassed several CEOs. How does it work? A fraud buys a painting—usually a colorful abstract work—from an amateur painter or art school student. Then he has it professionally framed. In fact the frame might be worth much more than the actual painting. The fraud presents himself as an artist's agent, and convinces the CEO or other senior managers that the artist is a master and the painting is valuable. Just what the company needs in its boardroom: a modern masterpiece!

The company buys the painting for a mere $56,000. It hangs in the boardroom. It may look dreadful, but the CEO assures everyone that it's a masterpiece. Then a company board member, who owns a small collection of good Canadian oils and prints, spots the painting. He notes that it's amateurish. Obviously the painter was only just learning to mix his paint. Who is this painter? The signature on the painting is illegible.

An appraiser arrives and confirms the truth. The matter is hushed up, after the CEO promises to leave art purchases to a knowledgeable committee and reputable dealers. Meanwhile, the fraud has disappeared with the money. He'll resurface months later in another city, along with a fresh supply of lousy abstract oils. This is a true story, and the fraudster remains active.

"It's amazing what a crook can get away with," says David Goldie. "Few of them are ever arrested and charged for selling bogus works. It's also amazing how lax people can be when they acquire art. It doesn't take much to prevent damage and loss. With a good security, careful handling, and a bit of luck, a piece could survive for millennia."

Sadly, that mixture of conditions isn't as common as it should be.

Post-traumatic stress: Finding a way to heal

40

Jack's grandfather returned from the battlefields of World War I with a row of medals and a case of shell shock. Jack (not his real name) remembers the old man breaking down in tears on the back porch of their Winnipeg family home.

"It happened often," says Jack, a 60-year-old adjuster in Victoria, British Columbia. "He'd just start crying. He told us that he couldn't stop thinking about what he saw as a young officer at Passchendaele. Most of his platoon was wiped out in a bombardment, and he never got over it. He had ugly flashbacks for the rest of his life."

Today's mental health specialists consider shell shock to be a form of post-traumatic stress disorder (PTSD), which they often diagnose in soldiers returning from war. But the disorder can afflict anyone who experiences a particularly disturbing event that overwhelms his or her consciousness. In the insurance sector, adjusters are the most commonly affected. Dealing with the aftermath of a regional disaster such as Hurricane Katrina or a major earthquake, adjusters see things that the rest of us can usually avoid.

What sorts of images lodge in the minds of adjusters in the field, and cause them overwhelming stress? There's no absolutely reliable guideline. What one adjuster can witness without pain will cause severe emotional damage to another.

"Trauma is not what happens to you, but the meaning you give it," says Lee Nicholas, a Vancouver psychotherapist who has dealt with numerous PTSD sufferers. "Two adjusters can experience exactly the same things in a disaster zone. One will come home and file his report and that's that. The other will fall apart, and require years of treatment."

Working in California after the 1989 Loma Prieta earthquake, Jack saw decaying bodies pulled from rubble. He watched buildings collapse and highways crumble. But what he mentions repeatedly are the dead dogs and cats that were left on roadsides. At that time, he didn't think that the sight of animal corpses would hit him so hard.

"When I got home, I felt fine," he says. "I liked my job and my family life was good. Then things started to change."

Jack felt constantly agitated, and had difficulty falling asleep. He began to have nightmares about dead animals. His wife warned him about his drinking, which had increased significantly since his post-quake experiences. They argued frequently.

"I didn't realize that I was in trouble until my family doctor told me that I needed help," he says. "In fact, there were a number of signs that I simply ignored. I thought that I could tough it out. After all, adjusters see all sorts of nasty things. You assume that you'll put these things out of your mind eventually. You don't expect them to stay with you."

Adjusters working in disaster zones can experience serious emotional problems. Fortunately, there are effective treatments.

Jack sought treatment from a psychiatrist, who diagnosed PTSD after a couple of hour-long appointments. The treatment, which lasted for 5 years, included antianxiety drugs and verbal therapy. Jack rarely has nightmares now, and he has stopped drinking. He has also stopped taking the prescription painkillers that he used to make himself drowsy. His relations with his wife have improved, and he feels fit enough to assume responsibility for new projects in his office. He was lucky to have a doctor who was familiar with the signs of PTSD.

"Not all health care providers know how to spot PTSD," says Nicholas. "It can manifest itself in many different ways."

Since sleep disruption is one of the most common signs of the disorder, Nicholas often begins his interviews with potential PTSD sufferers by asking how they're sleeping. He'll ask them to describe their dreams. Some clients will discuss repeated nightmares; others will talk about vivid and extremely disturbing daydreams that include visual, auditory, and olfactory sensations. Such daydreams can convince sufferers that they're losing their minds.

Self-medication becomes a crutch for many people with PTSD. Like Jack, they might start to drink heavily. Some will see abrupt changes in their drinking patterns, and might start using different drugs. Paul (not his real name), a Toronto-based independent adjuster, is an example. He doesn't want to talk about the fire and auto claims that he calls "messy," but he describes his substance abuse in detail.

"I figured that I was young and strong enough to handle the stuff I saw in the field," he says. "What I didn't realize was that my whole life had become a coping strategy. I popped pills and snorted cocaine. At the same time, I was taking on as much extra work as I could get. I stopped taking holidays, and didn't see my family for weeks on end."

Cocaine use led to a cardiac arrhythmia. Paul's GP referred him to a heart specialist. The specialist asked about Paul's stress level and work in the field. It was when Paul mentioned nightmares that the specialist became suspicious and sent him to a psychotherapist. Paul's arrhythmia is now under control, and he has spent the last 2 years undergoing psychotherapy for classic PTSD.

"For years I had all of the clinical criteria for the disorder," he says. "It took a heart condition to force me to deal with it."

In therapists' offices, PTSD sufferers emphasize their feelings of complete helplessness. Past emotional shocks impair their usual methods of dealing with troubling events. How might adjusters returning from post-Katrina New Orleans describe their feelings?

"If they've developed PTSD, they might tell me that they're numb," says Nicholas. "Or scared. They might feel constantly threatened and frightened. Whatever negative thoughts and feelings they have, they'll probably tell me that they can't shut them off. There's no relief, apart from behavior that will result in more problems, including addiction and workaholism."

Unfortunately, some may feel ashamed of taking their problem to a therapist. Especially among adjusters who believe that they can face any challenge, seeking help can involve embarrassment and shame. PTSD can result in forms of sexual dysfunction such as impotence—not the kind of weakness that many males are happy to discuss. There are also fears about one's professional image. If an adjuster can't handle the nasty side of work in disaster zones, what good is he or she? Naturally, that first visit to a therapist can be nerve-wracking.

"At first I didn't want to go too deeply into any subject with my psychiatrist," says Jack. "I just wanted enough pills to calm me down and stop me feeling so depressed all the time. I wasn't comfortable about talking about my relations with my wife. In fact, it took me months to start talking about my marriage."

Jack's psychiatrist coaxed him gently to discuss areas of his life that he had never talked about with anyone else. In part, the psychiatrist was looking for any predisposition that Jack might have had to PTSD. In many cases, what appears to be the triggering event is one of a series of incidents, each of which contributes to the development of a person's PTSD. Over 30 years as an adjuster, Jack has worked in numerous disaster zones, but he has also dealt with less catastrophic but nonetheless disturbing items.

For example, early in his career he handled auto claims. He'd inspect vehicles destroyed in collisions, and occasionally he'd notice bloodstains on dashboards and shattered windscreens. Years later, with the help of his psychiatrist, he recalled a highway accident in which a child had died. The evidence for the child's death had affected him deeply, although he didn't discuss the matter with anyone at the time.

"Perhaps that's where my PTSD started," says Jack. "I was carrying around that memory for ages, and I didn't know that it was setting me up for my emotional problems after Loma Prieta. But once I managed to identify it as something that I had to deal with, I felt a lot stronger."

Nicholas notes that there are different treatments for PTSD. No case is hopeless. Doctors might prefer a combination of drugs and psychotherapy, but he believes that there are more effective methods. One that is gaining acceptance is eye movement desensitization and reprocessing (EMDR), which allows sufferers to mentally reprocess the ways that shocking events affect their emotions. According to EMDR practitioners, it is possible to revise the meanings that we give to the things that frighten us.

On Jack's office wall in a simple frame hangs his grandfather's medals, including one for bravery. Jack points to the latter with pride.

"I like to imagine that they gave him that medal not for what he did in battle, but for what he had to live through when he came home. He didn't have the benefit of the kinds of treatments that are available to PTSD sufferers today. That's what I'd tell any adjuster who's having the nightmares and the depression. There's a way to heal."

Index

FSC
www.fsc.org
MIX
Papier | Fördert gute Waldnutzung
FSC® C083411

Kolibri 360 GmbH
Ferdinand-Jühlke-Straße 7
99095 Erfurt, Deutschland
produktsicherheit@kolibri360.de